Mandarin Chinese for Professionals Series:

Mando's Mandarin Chinese Course for Importing Construction Materials

By Zhou Hong

ISBN: 978-1-0674-3652-0

Published by Zhou Hong

Vancouver, Canada

First Edition

Website: www.cantomando.com

TABLE OF CONTENTS

ABOUT THE BOOK 5

ABOUT ME 8

ACKNOWLEDGEMENTS 9

LESSON 1: PRODUCTS AND SOURCING 11

LESSON 2: QUOTATION AND BILLING 26

LESSON 3: INSPECTION 49

LESSON 4: CUSTOMS AND SHIPPING 60

LESSON 5: BUILDING RELATIONSHIPS 75

FINAL PROJECT: YOUR PURCHASE ORDER 98

VOCABULARY LIST 101

关于本书

《普通话专题课程：建筑材料采购》是针对需要与中国伙伴和出口商高效沟通的建筑行业专业人士设计的普通话课程。本书每章包含大量与进口建筑材料相关的专业术语、与中国伙伴和出口商实际使用情境的丰富例句，以及与实际工作情况相匹配的清晰、以任务为导向的练习。

章节结构

- 第一课：产品与采购

学习产品规格、材料类别、供应商评估、工厂考察和采购流程的术语。实用对话涵盖询价邮件、产品对比和样品请求。

- 第二课：报价与账单

掌握报价、价格谈判、付款条款、国际贸易术语（Incoterms）、发票与成本分解等词汇。包括模板句用于请求报价、比较投标和确认订单。

- 第三课：检验

关注质量控制、检验标准、抽样、测试与报告。练习清晰表达检验标准、非合格记录与验收程序。

- 第四课：海关与运输

涵盖单据、清关程序、关税、海关编码（HS codes）、包装、运输条款和物流协调。提供用于海关查询、提单讨论和运输跟踪的实用脚本。

- 第五课：建立关系

强调长期伙伴关系、文化因素、谈判礼仪与售后支持。包括维护信任、解决问题和协同规划的策略。

这本高级中文教材适合建筑项目经理、采购人员、工程师、贸易专业人员以及希望与中国供应商合作或管理国际供应链的专业人士。学生可以用来自学或者作为课堂学习材料使用。

这本书能帮助学生建立在工作中所需的语言自信与实际操作能力，而下一步是将每章的词汇与真实的供应商联系结合起来，大声练习对话，并从询价到发货模拟一个完整的对话。

本书配套音频可在 CantoMando Courses 官方博客中收听，自学者可通过扫描下方二维码获取相关音频资源。

“行业精英普通话系列”还包括以下专题教材：《普通话专业课程：职场沟通》《普通话专业课程：时尚零售》《普通话专业课程：地产销售（上）》《普通话专业课程：地产销售（下）》《普通话专业课程：银行柜员》《普通话专业课程：建筑材料采购》《普通话专业课程：葡萄酒行业》等。

About the Book

Mando's Mandarin Chinese Course for Importing Construction Materials is designed for professional students in the construction industry who need to communicate efficiently with Chinese partners and exporters. It focuses on the vocabulary, expressions, and real-world patterns used in importing construction materials from China.

In this book, you'll learn to master field-specific terminology across key import stages, build fluency in professional conversations, negotiations, and written correspondence, and develop practical skills to source, quote, inspect, ship, and manage supplier relationships in Chinese.

Chapter breakdown

- Lesson 1: Products and Sourcing

Learn terminology for product specs, materials categories, supplier evaluation, factory visits, and sourcing workflows. Practical dialogues cover inquiry emails, product comparison, and sample requests.

- Lesson 2: Quotation and Billing

Gain vocabulary for quotations, price negotiations, payment terms, incoterms, invoices, and cost breakdowns. Includes model sentences for requesting quotes, comparing bids, and confirming orders.

- Lesson 3: Inspection

Focus on quality control, inspection standards, sampling, testing, and reporting. Practice clear communication for inspection criteria, non-conformance notes, and acceptance procedures.

- Lesson 4: Customs and Shipping

Cover documents, clearance procedures, duties, HS codes, packaging, shipping terms, and logistics coordination. Features practical scripts for customs inquiries, bill of lading discussions, and shipment tracking.

- Lesson 5: Building Relationships

Emphasizes long-term partnerships, cultural considerations, negotiation etiquette, and post-sale support. Includes strategies for maintaining trust, resolving issues, and collaborative planning.

Each chapter contains a large set of professional terminology tailored to importing construction materials; numerous example sentences illustrating real-world usage with Chinese partners and exporters and clear, task-oriented exercises that mirror industry scenarios. The book is suitable for self-study or classroom use, with emphasis on operational communication.

This book is an advanced level Chinese language textbook, it is perfect for construction project managers, procurement specialists, engineers, and trade professionals and students aiming to work with Chinese suppliers or manage international supply chains.

With this book, you'll build the language confidence and practical know-how to navigate the full lifecycle of importing construction materials from China. A practical next step is to pair each chapter's vocabulary with a real supplier contact, practice the dialogues aloud, and simulate a full end-to-end import conversation from inquiry to shipment.

The audio of the Chinese content of each chapter is available on the CantoMando Courses podcast, scan below to listen:

关于我

我是洪舟，但大部分学生都叫我 Jojo。我十三岁时从广州搬到加拿大温哥华，那是我第一次真正置身于一个完全不同的语言环境里。刚到那会儿，我每天都在努力听懂别人说什么，慢慢地，我发现语言不仅是沟通工具，更是让人和人彼此靠近的桥梁。

多年以后，当我还是一名自由翻译时，有位客户随口问了一句：“你教我中文吧！” 就是这句话，把我重新带回了语言的世界。从那之后，我开始教汉语，一教就是十多年。学生有来自各行各业的成年人，也有想和长辈沟通的海外华人年轻一代。我教过长者公寓里的爷爷奶奶，也教过奢侈品牌的销售顾问，甚至在2020年有机会加入温哥华社区学院（VCC）担任中文老师。

这些经历让我意识到：语言可以改变一个人，也可以改变一段关系。

而最有效的教学方式，就是为每一个学生量身打造学习路径。

与此同时，我也发现了一个长期存在的问题：中文教材不是太学术，就是太僵硬，很难真正符合学生的学习需求。于是我开始动手写自己的教材——灵活、有趣、能组合成各种课程的那种。

希望这套教材可以陪你走上一条轻松、有动力、真正“学得进去”的中文学习之路。也希望你能像我一样，在语言里发现人与人之间最温暖的联系。

About Me

My name is Jojo Hong, though most of my students simply call me Jojo. I moved from Guangzhou to Vancouver when I was thirteen, and found myself suddenly surrounded by English—a language I barely understood at the time. Every day felt like puzzle-solving, but slowly, I discovered something meaningful: language connects people in a way nothing else can.

Years later, while I was working as a freelance translator, a client casually said,
"Why don't you teach Chinese?"
That one sentence brought me back to the world of language—but this time, as a teacher.

Since then, I've spent over a decade teaching Mandarin and Cantonese to students from all kinds of backgrounds: adults learning for work, overseas Chinese reconnecting with their roots, or people simply curious about the language. I've taught seniors in retirement homes, sales professionals in luxury retail, and even served as a Mandarin instructor at Vancouver Community College in 2020.

Through these experiences, I learned that **language learning is most powerful when it's personal**. That's why I design my lessons with flexibility and individuality in mind.

Over the years, I realized that many existing textbooks were either too rigid or too academic, rarely matching what real learners needed. So I began writing my own teaching materials—books that teachers can mix and match, and students can use to guide their own journey.

I hope this series becomes a supportive companion in your learning process—something that brings clarity, confidence, and maybe even a little joy into your study of Chinese.

Acknowledgements

I want to thank my proofreader, Barbara Wu, you made this series possible.

Barbara, 感谢你付出和支持。

The audio for the Chinese content is available on the CantoMando Courses podcast. Please scan the QR code below to listen.

Lesson 1: Products and Sourcing

Vocabulary A:

- Supplier - 供应商 - **gōng yìng shāng**
- Manufacturer - 厂商 - **chǎng shāng**
- Original Manufacturer - 源头厂商 - **yuán tóu chǎng shāng**
- Wholesaler - 批发商 - **pī fā shāng**
- Factory - 工厂 - **gōng chǎng**
- Warehouse - 仓库 - **cāng kù**
- Exporter - 出口商 - **chū kǒu shāng**
- Construction Materials - 建筑材料 - **jiàn zhù cái liào**

Example:

Are you the original manufacturer? **你是源头厂商吗？** **Nǐ shì yuántóu chǎngshāng ma?**
We are the original manufacturer for doors and windows. **我们是门窗的源头厂商。** **Wǒmen shì ménchuāng de yuántóu chǎngshāng。**
Our suppliers are in Shandong province. **我们的供应商在山东省。** **Wǒmen de gōngyìngshāng zài Shāndōng shěng。**
I am an exporter of construction materials. **我是建筑材料的出口商。** **Wǒ shì jiànzhù cáiliào de chūkǒu shāng。**
Can I visit your factory? **我可以参观你们的工厂吗？** **Wǒ kěyǐ cānguān nǐmen de gōngchǎng ma?**
We have a warehouse in Guangdong province. **我们的仓库在广东省。** **Wǒmen de cāngkù zài Guǎngdōng shěng。**

Construction Materials - 建筑材料-**Jiàn zhù cái liào**

Steel **钢材** **gāng cái**	**H section** **H型钢** **H xíng gāng**	**C purlin** **C型檩条** **C xíng lǐn tiáo**	**Z purlin** **Z型檩条** **Z xíng lǐn tiáo**
Steel Rod **钢杆** **gāng gǎn**	**Angle Steel** **角钢** **jiǎo gāng**	**Gutter** **排水沟** **pái shuǐ gōu**	**Circular and Square Tube** **圆管和方管** **yuán guǎn hé fāng guǎn**
Steel Deck **楼承板 / 钢承板** **lóu chéng bǎn /gāng chéng bǎn**	**Single Steel Sheet** **单层钢板** **dān céng gāng bǎn**	**EPS Sandwich Panel** **EPS夹芯板** **EPS jiā xīn bǎn**	**Rockwool Sandwich Panel** **岩棉夹芯板** **yán mián jiā xīn bǎn**
PU Sandwich Panel **聚氨酯夹芯板** **jù ān zhǐ jiā xīn bǎn**	**Pulling Rivet** **拉铆钉** **lā mǎo dīng**	**Expansion Bolt** **膨胀螺栓** **péng zhàng luó shuān**	**High Strength Bolt** **高强度螺栓** **gāo qiáng dù luó shuān**
Foundation Bolt **基础螺栓** **jī chǔ luó shuān**	**Screw Nut** **螺母** **luó mǔ**	**Self Tapping Screw** **自攻螺钉** **zì gōng luó dīng**	**PVC Downspout** **PVC排水管** **PVC pái shuǐ guǎn**
Carbon Steel **碳钢** **tàn gāng**	**Door** **门** **mén**	**Window** **窗户** **chuāng hù**	**Galvanized Steel** **镀锌钢** **dù xīn gāng**
Stainless Steel **不锈钢** **bù xiù gāng**	**Aluminum Tube** **铝管** **lǚ guǎn**	**Slotted Track** **槽式导轨 / 带槽轨道** **cáo shì dǎo guǐ /dài cáo guǐ dào**	**Scaffolding** **脚手架** **jiǎo shǒu jià**
Cold Rolled Steel **冷轧钢** **lěng zhà gāng**			

- Size - 尺寸 - **chǐ cùn**
- Length - 长度 - **cháng dù**
- Width - 宽度 - **kuān dù**
- Depth - 深度 - **shēn dù**
- Thickness - 厚度 - **hòu dù**
- Diameter - 直径 - **zhí jìng**
- Metre - 米 - **mǐ**
- Centimetre - 厘米 - **lí mǐ**
- Millimeter -毫米 - **háo mǐ**
- KG - 公斤 - **gōng jīn**
- Ton - 吨 - **dūn**
- Weight - 重量 - **zhòng liàng**
- Load Capacity - 承重 - **chéng zhòng**
- Type - 型号 - **xíng hào**

Length:

number	Metres	Long
三 sān 3	米 mǐ metres	长 cháng long
九 jiǔ 9	米 mǐ metres	长 cháng long

Thickness:

number	MM	Thickness
210 èr bǎi yī shí	毫米 háo mǐ mm	厚 hòu thick
150 yī bǎi wǔ shí	毫米 háo mǐ mm	厚 hòu thick

Weight Or Load Capacity: :

number	KG	Weight
3 sān	公斤 gōng jīn Kg	重 zhòng heavy
1500 Yī qiān wǔ bǎi	公斤 gōng jīn Kg	重 zhòng heavy

Descriptions:

Description	de	Items
M6 bolt, 80 mm length M6号，80毫米长 M liù hào, bā shí háo mǐ cháng	的 de	Stainless Steel Bolt 不锈钢螺栓 bù xiù gāng luó shuān
40×40 40乘40 sì shí chéng sì shí	的 de	galvanized angle steel 镀锌角钢 dù xīn jiǎo gāng
1.7 m high × 0.95 m wide × 1.8 m long 1.7米高，0.95米宽，1.8米长 yī diǎn qī mǐ gāo, líng diǎn jiǔ wǔ mǐ kuān, yī diǎn bā mǐ cháng	的 de	Scaffolding 脚手架 jiǎo shǒu jià
6 metres 六米长 liù mǐ cháng	的 de	Aluminum square tube 铝合金方管 lǚ hé jīn fāng guǎn
2.4 mm diameter 直径2.4毫米 zhí jìng èr diǎn sì háo mǐ	的 de	Pull rivet 拉铆钉 lā mǎo dīng

Examples:

I want one thousand M6 stainless steel bolts with 80-mm length. **我要一千个M6号，80毫米长的不锈钢螺栓。** **wǒ yào yī qiān gè M liù hào, bā shí háo mǐ cháng de bù xiù gāng luó shuān。**
How much is the unit price of the 40×40 galvanized angle steel? **40乘40的镀锌角钢单价多少钱？** **sì shí chéng sì shí de dù xīn jiǎo gāng dān jià duō shǎo qián?**
How much weight can the scaffolding (1.7 m × 0.95 m × 1.8 m) support? **1.7米高，0.95米宽，1.8米长的手脚架最多能承重多少公斤？** **yī diǎn qī mǐ gāo, líng diǎn jiǔ wǔ mǐ kuān, yī diǎn bā mǐ cháng de shǒu jiǎo jià zuì duō néng chéng zhòng duō shǎo gōng jīn?**
Do you want the aluminum square tube in 4 metres or 6 metres? **你要四米长还是六米长的铝合金方管？** **nǐ yào sì mǐ cháng hái shì liù mǐ cháng de lǚ hé jīn fāng guǎn?**
What lengths are available for the pull rivets with 2.4-mm diameter? **直径二点四毫米的拉铆钉有哪些长度？** **zhí jìng èr diǎn sì háo mǐ de lā mǎo dīng yǒu nǎ xiē cháng dù?**

More Products:

Drywall **石膏板** **shí gāo bǎn**	**Shaftliner/Coreboard** **加固板** **jiā gù bǎn**	**Tape** **网格带** **wǎng gé dài**	**Drywall Adhesive** **石膏板胶水** **shí gāo bǎn jiāo shuǐ**
Plaster **灰泥** **huī ní**	**Putty** **腻子** **nì zi**	**Cement** **水泥** **shuǐ ní**	**Cement and Sand Mix** **水泥砂浆** **shuǐ ní shā jiāng**
Quickdry Plaster **快干石膏** **kuài gān shí gāo**	**Lime** **石灰** **shí huī**	**Texture Spray** **颗粒喷涂** **kē lì pēn tú**	**Vapour Barrier Membrane** **防潮隔气膜** **fáng cháo gé qì mó**
Rain Screen Wire **雨幕网** **yǔ mù wǎng**	**Water Barrier Membrane** **防水透气膜** **fáng shuǐ tòu qì mó**	**Stucco Corner** **护角网** **hù jiǎo wǎng**	**Bullnose Corner** **圆角护角条** **yuán jiǎo hù jiǎo tiáo**
Plaster Stop **石膏收边条** **shí gāo shōu biān tiáo**	**Wire** **铁丝** **tiě sī**	**Window Flashing** **窗户防水板** **chuāng hù fáng shuǐ bǎn**	**Tar Paper** **焦油纸** **jiāo yóu zhǐ**
Diamond Mesh **菱形网** **líng xíng wǎng**	**Galvanized Mesh** **镀锌网** **dù xīn wǎng**	**Strips** **条 / 板条** **tiáo /bǎn tiáo)**	**Styrofoam** **泡沫板** **pào mò bǎn**

Examples:

I would like to order 800 square meters of drywall. Do you have a 12mm thick dry wall? **我要订八百平方米的石膏板，你们有没有十二毫米厚的石膏板？** **Wǒ yào dìng bābǎi píngfāng mǐ de shígāobǎn, nǐmen yǒu méiyǒu shí'èr háomǐ hòu de shígāobǎn?**
Do you have a 150cm wide Water Barrier Membrane? What is the length in one roll? **你有没有一百五十厘米宽的防水透气膜？ 一卷有多长？** **Nǐ yǒu méiyǒu yībǎi wǔshí límǐ kuān de fángshuǐ tòuqìmó? Yī juǎn yǒu duō cháng?**
I need two hundred bags of quick dry plaster, when can you deliver? **我要两百袋快干水泥，你什么时候能送货？** **Wǒ yào liǎngbǎi dài kuàigān shuǐní, nǐ shénme shíhòu néng sònghuò?**
Our aluminum rain screen wire comes in three thickness: 1mm, 1.2mm and 1.5mm. **我们的铝合金雨幕网有三种厚度，一毫米，一点二毫米和一点五毫米。** **Wǒmen de lǚhéjīn yǔmùwǎng yǒu sān zhǒng hòudù: yī háomǐ, yīdiǎn'èr háomǐ hé yīdiǎn'wǔ háomǐ。**

- Materials - 材料- **cáiliào**

English	Chinese	Example
Expensive	贵 **guì**	这种材料很贵！ **Zhè zhǒng cáiliào hěn guì!**
Cheap	便宜 **piányí**	你们有没有更便宜一点的材料？ **Nǐmen yǒu méiyǒu gèng piányí yìdiǎn de cáiliào?**
Light	轻 **qīng**	这种材料很轻。 **Zhè zhǒng cáiliào hěn qīng。**
Insulation	保温材料 / 隔热材料 **bǎowēn cáiliào / gé rè cáiliào**	这种夹芯板有绝缘材料。 **Zhè zhǒng jiāxīnbǎn yǒu juéyuán cáiliào。**
Water Resistant	防水 **fángshuǐ**	这种膜是防水的。 **Zhè zhǒng mó shì fángshuǐ de。**
Fire Resistant	防火 **fánghuǒ**	这是防火的夹芯板。 **Zhè shì fánghuǒ de jiāxīnbǎn。**
Moisture Resistant	防潮 **fángcháo**	这种板材是防潮的。 **Zhè zhǒng bǎncái shì fángcháo de。**
Sound Proof	隔音 **géyīn**	我要一种隔音的天花板。 **Wǒ yào yī zhǒng géyīn de tiānhuābǎn。**
Wood	木 **mù**	这是木的门窗。 **Zhè shì mù de ménchuāng。**

Translate the following sentences:

- Where is your factory located?
- Are you the original manufacturer or a supplier?
- What materials are used in the EPS sandwich panel?
- What is the width of the water barrier membrane?

- Do you have aluminum square tubes? What is the diameter?
- When can you deliver the quick dry plaster?
- Do you have aluminum rain screen wire in different thicknesses?
- What is the difference between self-tapping screws and pull rivets?

Answer:

- Where is your factory located?
- 你们的工厂在哪里？
- **Nǐmen de gōngchǎng zài nǎlǐ?**

- Are you the original manufacturer or a supplier?
- 你们是源头厂商还是供应商？
- **Nǐmen shì yuántóu chǎnshāng háishì gōngyìngshāng?**

- What materials are used in the EPS sandwich panel?
- EPS夹芯板有什么材料？
- **EPS jiāxīn bǎn yǒu shénme cáiliào?**

- What is the width of the water barrier membrane?
- 防水膜的宽度是多少？
- **Fángshuǐ mó de kuāndù shì duōshǎo?**

- Do you have aluminum square tubes? What is the diameter?
- 你们有铝合金方管吗？边长是多少？
- **Nǐmen yǒu lǚhéjīn fāngguǎn ma? Biāncháng shì duōshǎo?**

- When can you deliver the quick dry plaster?
- 快干石膏什么时候可以交货？
- **Kuàigān shígāo shénme shíhòu kěyǐ jiāohuò?**

- Do you have aluminum rain screen wire in different thicknesses?
- 你们有不同厚度的铝合金雨幕网吗？
- **Nǐmen yǒu bùtóng hòudù de lǚhéjīn yǔmù wǎng ma?**

- What is the difference between self-tapping screws and pull rivets?
- 自攻螺钉和拉铆钉有什么区别？
- **Zìgōng luódīng hé lā mǎodīng yǒu shénme qūbié?**

Make a list of your shopping list, including the sizes and quantity.

Material	Size	Quantity	Desirable Features
Example: 石膏板 **Example: shí gāo bǎn**	**10毫米厚** **shí háo mǐ hòu**	**一千平方米** **yì qiān píng fāng mǐ**	**防水透气** **fáng shuǐ tòu qì**

You need to find a supplier for steel frames for a large construction project. What questions would you ask?

Question 1	
Question 2	
Question 3	
Question 4	
Question 5	
Question 6	

You are a drywall manufacturer in China. How do you answer the following questions from your clients?

Question 1	**石膏板有哪些不同的尺寸？** **shí gāo bǎn yǒu nǎ xiē bù tóng de chǐcùn?**
Answer	
Question 2	**最厚的石膏板有多厚？** **zuì hòu de shí gāo bǎn yǒu duō hòu?**
Answer	
Question 3	**有没有保温隔热的石膏板？** **yǒu méi yǒu bǎo wēn gé rè de shígāobǎn?**
Answer	
Question 4	**你们有夹芯板吗？ 有没有防水的夹芯板？** **nǐ men yǒu jiā xīn bǎn ma? yǒu méi yǒu fáng shuǐ de jiāxīnbǎn?**
Answer	
Question 5	**你们有隔音的材料吗？** **nǐ men yǒu gé yīn de cáiliào ma?**
Answer	
Question 6	**你们有哪几种不同的石膏板？** **nǐmen yǒu nǎ jǐ zhǒng bù tóng de shígāobǎn?**
Answer	

Lesson 2: Quotation and Billing

Vocabulary A:

- Quote - 报价 - **bào jià**
- Exchange Rate - 汇率 - **huì lǜ**
- MOQ - 最小起订量 - **zuì xiǎo qǐ dìng liàng**
- Ready Stock - 现货 - **xiàn huò**
- Free On Board - 船上交货价 - **chuán shàng jiāo huò jià**
- C&F - 成本加运费 - **chéng běn jiā yùn fèi**
- CIF (Cost Insurance Freight) - 成本加保险加运费 - **chéng běn jiā bǎo xiǎn jiā yùn fèi**
- DDP - 完税后交货 - **wán shuì hòu jiāo huò**
- EXW - 出厂价 - **chū chǎng jià**
- Purchase Order - 采购订单 - **cǎi gòu dìng dān**
- Proforma Invoice - 形式发票 - **xíng shì fā piào**
- Commercial Invoice - 商业发票 - **shāng yè fā piào**
- Contract - 合同 - **hé tóng**
- Letter of Credit (L/C) - 信用证 - **xìn yòng zhèng**
- Telegraphic Transfer (T/T) - 电汇 - **diàn huì**
- Documents Against Payment (D/P) - 付款交单 - **fù kuǎn jiāo dān**
- Payment Method - 付款方式 - **fù kuǎn fāng shì**
- Deposit - 定金 - **dìng jīn**

Quote - 报价 - **bào jià**

- Unit Price - 单价 - **dān jià**
- Quantity - 数量 - **shù liàng**
- Total - 总价 - **zǒng jià**
- Shipping - 运输 - **yùn shū**
- Insurance - 保险 - **bǎo xiǎn**
- Shipment Value - 货物价值 - **huò wù jià zhí**
- Customs and Duties - 关税 - **guān shuì**
- Quality Inspection Fee - 检验费 - **jiǎn yàn fèi**

Examples of Quotation:

	Product Name 产品名称 chǎn pǐn míng chēng	Specification 规格 guī gé	Quantity 数量 shù liàng	Unit Price (USD) 单价 （USD) dān jià (USD)	Total (USD) 总价 (USD) zǒng jià (USD)
1	Galvanized Square Tube 镀锌方型管 dù xīn fāng xíng guǎn	6 meters 六米 liù mǐ	100 tons 100吨 yī bǎi dūn	500 USD/ton 500元/吨 wǔ bǎi yuán/dūn	50,000 50000 wǔ wàn
2	Galvanized Square Tube 镀锌方管 dù xīn fāng guǎn	9 meters 九米 jiǔ mǐ	200 tons 200吨 èr bǎi dūn	800 USD/ton 800元/吨 bā bǎi yuán/dūn	160,000 160,000 shí liù wàn
				Total 合计 hé jì	210,000 210,000 èr shí yī wàn

Notes
备注：
bèi zhù

1. **This quote includes packaging and shipping fees.**
   本报价包含包装费和运输费。
   Běn bào jià bāo hán bāo zhuāng fèi hé yùn shū fèi。

2. **Payment method: 100% advance payment, goods shipped upon receipt of payment.**
   付款方式：100%预付，收到货款后发货。
   Fù kuǎn fāng shì: 100% yù fù, shōu dào huòkuǎn hòu fāhuò。

3. **Validity: The quote is valid for 30 days from the quotation date.**
   报价有效期： 报价日期起三十天内。
   Bào jià yǒu xiào qī: bào jià rì qī qǐ sān shí tiān nèi。

Examples:

<table>
<tr><th></th><th>Product Name
产品名称
chǎn pǐn míng chēng</th><th>Specification
规格
guī gé</th><th>Quantity
数量
shù liàng</th><th>Unit Price (USD)
单价 （USD)
dān jià (USD)</th><th>Total (USD)
总价 (USD)
zǒng jià (USD)</th></tr>
<tr><td>1</td><td>304 Stainless Steel Screw
304不锈钢螺丝
304 bùxiùgāng luósī</td><td>M6
M6
M6</td><td>30000
30000
sān wàn</td><td>0.5
0.5
líng diǎn wǔ</td><td>15,000
15,000
yī wàn wǔ qiān</td></tr>
<tr><td>2</td><td>304 Stainless Steel Nut
304不锈钢螺母
304 bù xiù gāng luó mǔ</td><td>M6
M6
M6</td><td>30000
30000
sān wàn</td><td>0.1
0.1
líng diǎn yī</td><td>3,000
3,000
sān qiān</td></tr>
<tr><td colspan="4" rowspan="4">1. The above quote includes tax, with a VAT rate of 10%.
以上报价含税，增值税的税率为10%。
Yǐ shàng bào jià hán shuì, zēng zhí shuì de shuì lǜ wéi bǎifēnzhī shí。

2. The quote is valid for 30 days.
报价有效期是三十天。
Bào jià yǒu xiào qī shì sān shí tiān。</td><td>Subtotal
小计
xiǎo jì</td><td>18,000
yī wàn bā qiān</td></tr>
<tr><td>Tax Rate
税率
shuì lǜ</td><td>10%
bǎi fēn zhī shí</td></tr>
<tr><td>Sales Tax
销售税
xiāo shòu shuì</td><td>1,800
yī qiān bā bǎi</td></tr>
<tr><td>Total
合计
hé jì</td><td>19,800
yī wàn jiǔ qiān bā bǎi</td></tr>
</table>

Delivery Time:
交货期：
Jiāo huò qī :

Within one week after signing the contract
签订合同后一周内

qiān dìng hé tóng hòu yī zhōu nèi

Delivery Method: Door-to-door delivery

发货方式：送货上门

Fā huò fāng shì: sòng huò shàng mén

Payment Method: Telegraphic Transfer (T/T) or Documents Against Payment (D/P)
付款方式：电汇或付款交单

Fù kuǎn fāng shì: diàn huì huò fù kuǎn jiāo dān

Example Price List:

Item Number 货号 huò hào	**Y001**	**Y002**
Product 产品 chǎn pǐn	**Grey Aluminum Door Frame 灰色铝合金门框 huī sè lǚ hé jīn mén kuàng**	**White Aluminum Cupboard Door Panels 白色铝合金柜门板 bái sè lǚ hé jīn guì mén bǎn**
Specification 产品规格 chǎn pǐn guī gé	**24cm èr shí sì lí mǐ**	**60cm x 80cm liù shí lí mǐ x bā shí lí mǐ**
FOB 离岸价 lí àn jià	**200USD èr bǎi měi yuán**	**30USD sān shí měi yuán**
QTY/CTN PCS 每箱件数 měi xiāng jiàn shù	**50 wǔ shí**	**100 yī bǎi**
CTN Measures 外箱尺寸 wài xiāng chǐ cùn	**1m x 1m x 1m yī mǐ x yī mǐ x yī mǐ**	**1m x 1m x 1m yī mǐ x yī mǐ x yī mǐ**
Net Weight 产品净重 chǎn pǐn jìng zhòng	**1 ton yī dūn**	**1 ton yī dūn**
G W 产品毛重 chǎn pǐn máo zhòng	**1.5 ton yī diǎn wǔ dūn**	**1.5 tons yī diǎn wǔ dūn**

Other specifications:

Single Package Type, Material and Size	**单个包装的类型，材料和尺寸** **Dān gè bāo zhuāng de lèi xíng, cái liào hé chǐ cùn**
Inner Package's Type, Material and Size	**内包装的类型，材料和尺寸** **Nèi bāo zhuāng de lèi xíng, cái liào hé chǐ cùn**
Quantity/ Inner Package	**内包装的产品数量** **Nèi bāo zhuāng de chǎn pǐn shù liàng**
CTN/20'; QTY/20'	**每个20 '柜的箱数；每个20' 柜的产品数量** **Měi gè 20' guì de xiāng shù; měi gè 20' guì de chǎn pǐn shù liàng**
CTNs/40'; QTY/40'	**每个40 '柜的箱数；每个40' 柜的产品数量** **Měi gè 40' guì de xiāng shù; měi gè 40' guì de chǎn pǐn shù liàng**
Delivery Time	**交货期** **Jiāo huò qī**
Other Terms	**其他条款** **Qí tā tiáo kuǎn**

Example Price List

Product Code 产品编号 chǎn pǐn biān hào	Product Name 产品名称 chǎn pǐn míng chēng	Product Specification 产品参数 chǎn pǐn cān shù	Size 尺寸 chǐ cùn	Package 包装 bāo zhuāng	MOQ 最小起订量 zuì xiǎo qǐ dìng liàng	Unit Price 单价 dān jià	Unit/Carton 每箱数量 Měi xiāng shù liàng
A0001	**Scaffolding 脚手架 jiǎo shǒu jià**	**6M**	**6M x 6M**	**Pallet 托盘 tuō pán**	**1,000 yī qiān**	**20 USD èr shí měi yuán**	**500**
A0002	**Scaffolding Ladder 脚手架梯子 jiǎo shǒu jià tī zi**	**2M**	**2M**	**Pallet 托盘 tuō pán**	**500**	**15 USD shí wǔ měi yuán**	**1,000**

Lead Time	交货期 jiāo huò qī
20-40 Days, subject to the product and the quantity you order. 20-40天，取决于你订购的产品和数量。 20-40 tiān, qǔ jué yú nǐ dìng gòu de chǎn pǐn hé shù liàng。	
Payment Terms	付款条款
T/T or L/C 电汇或者信用证。 Diàn huì huò zhě xìn yòng zhèng。	
Minimum Order Quantity	最小起订量
5,000 USD per order, there is a 300 USD handling fee for orders less than 5,000 USD. 每单最少5,000美元，低于5,000美元的订单要收取300美元的手续费。 Měi dān zuì shǎo 5,000 měi yuán, dī yú 5,000 měi yuán de dìng dān yào shōu qǔ 300 měi yuán de shǒuxù fèi。	
Sample Policy	样品政策
Products under 100 USD are free. 100 美元以下的样品免费。 100 měi yuán yǐ xià de yàng pǐn miǎn fèi。	

Cost Breakdown - 成本明细 - **chéng běn míng xì**

1. **Cost of stainless steel frames:**

   不锈钢框架的价钱

   bù xiù gāng kuàng jià de jià qián

   - **Unit Price: $2,000 per ton**
   - **单价： 每吨$2000**
   - **dān jià: měi dūn $2,000**
   - **Quantity: 100 tons**
   - **数量：一百吨**
   - **shù liàng: yī bǎi dūn**
   - **Total Cost: 100 tons x $2,000/ton = $200,000**
   - **总价：100 tons x $2,000/ton = $200,000**
   - **zǒng jià: yī bǎi dūn x $2,000 / dūn = $200,000**

2. **Shipping and Handling:**

   运输和手续费：

   yùn shū hé shǒu xù fèi:

   - **If shipping overseas, cost may vary depending on distance, method, and weight.**
   - **海外运输费取决于距离、运输方式和重量。**
   - **Hǎi wài yùn shū fèi qǔ jué yú jù lí, yùn shū fāng shì hé zhòng liàng。**
   - **Shipping: $5,000**
   - **运费：$5000**

- yùn fèi: $5,000

3. Insurance:

保险：

bǎo xiǎn:

- **Insurance costs for transportation often run around 1-3% of the total shipment value.**
- **运输保险费一般是货物总价的1-3%.**
- **Yùn shū bǎo xiǎn fèi yī bān shì huò wù zǒng jià de 1-3%。**
- **Example: If total goods value is $200,000, 2% insurance would be: $200,000 x 0.02 = $4,000**
- **比如：如果货物总价值是$200,000，2%的保险费是：$200,000 x 0.02 = $4,000**
- **Bǐ rú: rú guǒ huò wù zǒng jià zhí shì $200,000, 2% de bǎo xiǎn fèi shì: $200,000 x 0.02 = $4,000**

4. Customs and Duties:

关税及其他税费：

guān shuì:

- **Depending on the destination country, customs duties may apply.**
- **目的地国家可能会征收不同的关税。**
- **Mù dì dì guó jiā kě néng huì zhēng shōu bù tóng de guān shuì。**

5. Quality Inspection Fee :

检验费用：

jiǎn yàn fèi yòng :

- **If an inspection is needed before shipping, include this cost.**
- **如果运输前需要检验，请计入此项费用。**
- **Rú guǒ yùn shū qián xū yào jiǎn yàn, qǐng jì rù cǐ xiàng fèi yòng。**
- **Export Inspection: $500**
- **出口检验: $500**
- **chū kǒu jiǎn yàn: $500**
- Proforma Invoice - 形式发票 - **xíng shì fā piào**

（Seller - 卖方 - mài fāng)

ABC Construction Supplies

123 Building Ave

Cityville, ST 12345

Phone: (123) 456-7890

电话：

diàn huà:

Email: sales@abcconstruction.com

邮箱：

yóu xiāng:

(Buyer - 买方 - mǎifāng)

XYZ Construction Company

456 Contractor Blvd

Townsville, ST 67890

Phone: (987) 654-3210

电话：

diàn huà:

Email: info@xyzconstruction.com

邮箱：

yóu xiāng:

Invoice Number: PI-001

发票号码：

fā piào hào mǎ:

Invoice Date: October 3, 2023

日期：

rì qī:

Due Date: October 17, 2023

截止日期：

jié zhǐ rì qī:

Product 产品 chǎn pǐn	Quantity 数量	Unit Price 单价	Total 总价
Cement (50KG Bags) 水泥 （五十公斤袋） shuǐ ní (wǔ shí gōng jīn dài)	100	$5.00	$500.00
Gravel (ton) 建筑碎石（吨） jiàn zhù suì shí (dūn)	10	$50.00	$500.00
Subtotal 小计 xiǎo jì	$1,000.00		
Tax (10%) 税 shuì (bǎi fēn zhī shí)	$100.00		
Total Amount: 总计： zǒng jì	$1,100.00		

Payment Terms: Payment due within 14 days from the invoice date.

付款条款： 在发票日期的十四天内付款。

fù kuǎn tiáo kuǎn: Zài fā piào rì qī de shí sì tiān nèi fù kuǎn。

Thank you for your business! Please reference the invoice number when making the payment.

谢谢惠顾！付款时请提供发票号码。
xiè xiè huì gù! fù kuǎn shí qǐng tí gōng fā piào hào mǎ。

Seller Signature:____________________

卖方签名：

mài fāng qiān míng:

Position: Sales Manager

职位：销售经理

zhí wèi: xiāo shòu jīng lǐ

This is not a commercial invoice.

这不是商业发票。

zhè bù shì shāng yè fā piào。

- Commercial Invoice - 商业发票 - **shāng yè fā piào**

（Seller - 卖方 - mài fāng)

ABC Construction Supplies

123 Building Ave

Cityville, ST 12345

Phone: (123) 456-7890

电话：

diàn huà:

Email: sales@abcconstruction.com

邮箱：

yóu xiāng:

(Buyer - 买方 - mǎi fāng)

XYZ Construction Company

456 Contractor Blvd

Townsville, ST 67890

Phone: (987) 654-3210

电话：

diàn huà:

Email: info@xyzconstruction.com

邮箱：

yóu xiāng:

Invoice Number: PI-001

发票号码：

fā piào hào mǎ:

Invoice Date: October 3, 2023

日期：

rì qī:

Due Date: October 17, 2023

截止日期：

jié zhǐ rì qī:

Country of Origin

发件地国家：

fā jiàn dì guó jiā ：

Country of Destination

目的地国家：

mù dì dì guó jiā ：

Shipping Method: Ground Shipping

运输： 陆运
yùnshū：Lùyùn

Terms of Sale: FOB - Free On Board

付款条款：离岸价

fùkuǎn tiáokuǎn：Lí'ànjià

Consignment Number:

运单号：

yùn dān hào:

Exporter References:

出口商编号：

chūkǒu shāng biānhào：

Shipping Company:

运输公司：

yùn shū gōng sī:

Insurance Certification Number:

投保单号：

tóu bǎo dān hào:

Insured Value:

投保金额：

tóu bǎo jīn é:

Currency:

币种：

bì zhǒng :

Product **产品** **chǎn pǐn**	**Quantity** **数量** **shù liàng**	**Unit Price** **单价** **dān jià**	**Total** **总价** **zǒng jià**
Cement (50KG Bags) **水泥 （五十公斤袋）** **shuǐ ní (wǔ shí gōng jīn dài)**	**100**	**$5.00**	**$500.00**
Gravel (ton) **建筑碎石（吨）** **jiàn zhù suì shí (dūn)**	**10**	**$50.00**	**$500.00**
Destination City **目的地城市** **mù dì dì chéng shì**	**Origin of Goods** **原产地** **yuán chǎn dì**	**Net Mass** **净重** **jìng zhòng**	**Gross Mass** **毛重** **máo zhòng**
Subtotal **小计** **xiǎo jì**	**$1,000.00**		
Shipping Charge **运费** **yùn fèi**	**$100.00**		
Total Amount: **总计：** **zǒng jì**	**$1,100.00**		

Payment Instructions:

付款说明：

fù kuǎn shuō míng:

Please make payment to:

请付款到以下账户：

qǐng fù kuǎn dào yǐ xià zhàng hù:

ABC Construction Supplies

Bank Name: Builder's Bank

银行：

yín háng:

Account Number: 123456789

账户号码：

zhàng hù hào mǎ:

Routing Number: 987654321

路由号码：

lù yóu hào mǎ:

Thank you for your business! Please reference the invoice number when making the payment.

谢谢惠顾！付款时请提供发票号码。

xiè xiè huì gù! fù kuǎn shí qǐng tí gōng fā piào hào mǎ。

Seller Signature:____________________

卖方签名：

mài fāng qiān míng:

Position: Sales Manager

职位：销售经理

zhí wèi: xiāo shòu jīng lǐ

For Customs Clearance

用于清关

yòng yú qīng guān

- Packing List - 装箱单 - **zhuāng xiāng dān**

Example:

<table>
<tr><td>To
致
zhì</td><td colspan="2"></td><td>L/C Number
信用证号
xìn yòng zhèng hào</td><td colspan="3"></td></tr>
<tr><td colspan="3" rowspan="3">Importer Name
进口商名字
jìn kǒu shāng míng zì

Address
地址
dì zhǐ

Contact Info
联系方式
lián xì fāng shì</td><td>Invoice Number
发票号码
fā piào hào mǎ</td><td colspan="3"></td></tr>
<tr><td>Date
日期
rì qī</td><td colspan="3"></td></tr>
<tr><td>Contract Number
合同号码
hé tóng hào mǎ</td><td colspan="3"></td></tr>
<tr><td>Mark and Numbers
唛头和号码
mài tóu hé hào mǎ</td><td>Goods Descriptio n
货物描述
huòwù miáoshù</td><td>CTNS
件数
jiàn shù</td><td>Quantity
数量
shù liàng</td><td>Gross Weight
毛重
máo zhòng</td><td>Net Weight
净重
jìng zhòng</td><td>Measurement
体积
tǐ jī</td></tr>
<tr><td></td><td></td><td></td><td></td><td></td><td></td><td></td></tr>
<tr><td></td><td>Total
合计
hé jì</td><td>Total CTN
（总件数）
zǒng jiàn shù</td><td>Total Quantity
（总数量）
zǒng shù liàng</td><td>Total Gross Weight
（总毛重）
zǒng máo zhòng</td><td>Total Net Weight
（总净重）
zǒng jìng zhòng</td><td>Total Measurement
（总体积）
zǒng tǐ jī</td></tr>
<tr><td colspan="7">Total Packed in 10 CTNS ONLY
出货总件数十箱
chū huò zǒng jiàn shù shí xiāng</td></tr>
</table>

Vocabulary B:

- Discount - 折扣 - **zhé kòu**
- Fee - 费用 - **fèi yòng**
- Detailed - 详细的 - **xiáng xì de**
- Lead Time - 交货时间 - **jiāo huò shí jiān**
- Samples - 样品 - **yàng pǐn**
- Quality Control - 质量控制 - **zhì liàng kòng zhì**
- Production - 生产 - **shēng chǎn**
- Return Policy - 退货政策 - **tuì huò zhèng cè**
- Return(goods) - 退货 - **tuì huò**
- Standard - 标准 - **biāo zhǔn**
- Meet Standard - 符合标准 - **fú hé biāo zhǔn**
- Large Orders - 大额订单 - **dà'é dìng dān**
- Flexible - 灵活 - **líng huó**
- Change - 改变 - **gǎi biàn**
- Product Specs - 产品规格 - **chǎn pǐn guī gé**
- Catalog - 产品目录 - **chǎn pǐn mù lù**
- Competitor - 竞争者 - **jìng zhēng zhě**
- Defective - 有瑕疵的 - **yǒu xiácī de**
- Warranty - 保修期 - **bǎo xiū qī**
- Damaged - 损坏的 - **sǔn huài de**

- Inspection Reports - 检验报告 - **jiǎn yàn bào gào**

Common questions to ask the supplier:

1	**Can you give a better price?**
Chinese	**可以打折吗？** **kě yǐ dǎ zhé ma?**
2	**Can you give a 5% discount?**
Chinese	**你能给5%的折扣吗？** **nǐ néng gěi bǎi fēn zhī wǔ de zhē kòu ma?**
3	**Can you give a detailed price breakdown for the quote, including tax and fees?**
Chinese	**你能提供详细的包括税金和费用的价格明细吗？** **nǐ néng tí gōng xiáng xì de bāo kuò shuì jīn hé fèi yòng de jià gé míng xì ma?**
4	**What is the MOQ for this product?**
Chinese	**这个产品的最低起订量是多少？** **zhè gè chǎn pǐn de zuì dī qǐ dìng liàng shì duō shǎo?**
5	**What is the estimated lead time for this product?**
Chinese	**这个产品的交货时间大概要多久？** **zhè gè chǎn pǐn de jiāo huò shí jiān dà gài yào duō jiǔ?**
6	**Can you send samples?**
Chinese	**你们可以寄样品吗？** **nǐ men kě yǐ jì yàng pǐn ma?**
7	**What kind of quality control do you have for this product?**
Chinese	**这个产品有哪些产品控制措施？** **zhè gè chǎn pǐn yǒu nǎ xiē chǎn pǐn kòng zhì cuò shī?**
8	**Do you have a return policy for products that don't meet our standards?**
Chinese	**你们有退货政策吗？不符合我们标准的产品可以退货吗？** **nǐ men yǒu tuì huò zhèng cè ma? bù fú hé wǒ men biāo zhǔn de chǎn pǐn kě yǐ tuì huò ma?**
9	**Will you help us with customs clearance, or are we responsible for customs clearance?**
Chinese	**你会帮我们清关吗？还是我们自己负责清关？** **nǐ huì bāng wǒ men qīng guān ma? hái shì wǒ men zì jǐ fù zé qīng guān?**
10	**Are there discounts for large orders?**

Chinese	大额订单可以打折吗？ dà é dìng dān kě yǐ dǎ zhé ma?
11	How flexible are you with order changes after the initial order is placed?
Chinese	下单以后还能灵活地更改订单吗？ xià dān yǐ hòu hái néng líng huó de gēng gǎi dìng dān ma?
12	We need product specs in PDF format.
Chinese	我们需要PDF格式的产品规格表。 wǒ men xū yào PDF gé shì de chǎn pǐn guī gé biǎo。
13	How soon can you ship after payment?
Chinese	付款后多久能发货？ fù kuǎn hòu duō jiǔ néng fā huò?
14	Send me your latest catalog.
Chinese	请发你们最新的产品目录给我。 qǐng fā nǐ men zuì xīn de chǎn pǐn mù lù gěi wǒ。
15	Can you match this competitor's price?
Chinese	你们能给出这个竞争者的价格吗？ nǐ men néng gěi chū zhè gè jìng zhēng zhě de jià gé ma?
16	We prefer flexible payment terms.
Chinese	我们想要更灵活的付款方式。 wǒ men xiǎng yào gèng líng huó de fù kuǎn fāng shì。
17	Is there a warranty period?
Chinese	有没有保修期？ yǒu méi yǒu bǎo xiū qī?
18	We need faster shipping options.
Chinese	我们需要更快的运输方式。 wǒ men xū yào gèng kuài de yùn shū fāng shì。
19	Our last shipment arrived damaged.
Chinese	上一批货到的时候有损坏。 shàng yī pī huò dào de shí hòu yǒu sǔn huài。
20	Your packaging needs improvement.

Chinese	你们的包装得改进一下。 nǐ men de bāo zhuāng děi gǎi jìn yī xià。
21	Can you extend the payment deadline?
Chinese	付款期限能延后吗？ fù kuǎn qī xiàn néng yán hòu ma?
22	Your courier charges are too high.
Chinese	你们的快递费用太高了。 nǐ men de kuài dì fèi yòng tài gāo le。
23	Do you have local warehouses?
Chinese	你们有本地仓库吗？ nǐ men yǒu běn dì cāng kù ma?
24	We found cheaper alternatives.
Chinese	我们找到更便宜的替代品了。 wǒ men zhǎo dào gèng pián yí de tì dài pǐn le。
25	Your customs paperwork is incomplete.
Chinese	你们的清关文件不全。 nǐ men de qīng guān wén jiàn bù quán。
26	We need bilingual packaging labels.
Chinese	我们需要双语包装标签。 wǒ men xū yào shuāng yǔ bāo zhuāng biāo qiān。
27	Can you reduce the packaging size?
Chinese	能缩小包装体积吗？ néng suō xiǎo bāo zhuāng tǐ jī ma?
28	Can you guarantee delivery dates?
Chinese	你能保证交货日期吗？ nǐ néng bǎo zhèng jiāo huò rì qī ma?

29	**Can you shorten lead time?**
Chinese	**你能缩短生产周期吗？** **nǐ néng suō duǎn shēng chǎn zhōu qī ma?**
30	**The samples arrived crushed.**
Chinese	**样品寄过来都压坏了。** **yàng pǐn jì guò lái dōu yā huài le。**

Lesson 3: Inspection

Vocabulary A:

- Inspection - 检验 - **jiǎn yàn**
- Pre-Shipment Inspection - 装运前检验 - **zhuāng yùn qián jiǎn yàn**
- Export Inspection - 出口商检 - **chū kǒu shāng jiǎn**
- Quality Inspection - 质检 - **zhì jiǎn**
- Third Party Inspection 第三方检验 - **dì sān fāng jiǎn yàn**
- Pre-Production Inspection - 产前检验 - **chǎn qián jiǎn yàn**
- During-Production Inspection - 生产中检验 - **shēng chǎn zhōng jiǎn yàn**
- Final Inspection - 最终检验 - **zuì zhōng jiǎn yàn**
- Container Loading Inspection - 装柜监装 - **zhuāng guì jiān zhuāng**
- Random Inspection - 随机抽检 - **suí jī chōu jiǎn**
- Re-inspection - 复检 - **fù jiǎn**

Examples:

The client requires a third party inspection. **客户要求第三方检验。** **kè hù yāo qiú dì sān fāng jiǎn yàn。**
The manufacturer completed the export inspection. **厂家完成了出口商检。** **chǎng jiā wán chéng le chū kǒu shāng jiǎn。**
The inspector will do a random inspection on the goods. **检验员会随机抽查货物。** **jiǎn yàn yuán huì suí jī chōu chá huò wù。**
Did you do a pre-shipment inspection? **你有没有做装运前检验?** **nǐ yǒu méi yǒu zuò zhuāng yùn qián jiǎn yàn。**
The final inspection will be completed this week. **最终检验这个星期会完成。** **zuì zhōng jiǎn yàn zhè gè xīng qī huì wán chéng。**

- Product Certification - 产品认证 - **chǎn pǐn rèn zhèng**

建筑材料常用的国际认证：

jiàn zhù cái liào cháng yòng de guó jì rèn zhèng

CE	**FCC**	**ETL**	**SAA**
PSE	**UL**	**COC**	**ISO9000**
CCC	**UKCA**	**CSA**	**CCMC**
ICC	**EAC**		

Examples:

Is this product CE certified? **这个产品有没有CE认证？** **zhè gè chǎn pǐn yǒu méi yǒu CE rèn zhèng?**
The goods need to be CSA certified. **货物一定要有CSA认证。** **huò wù yí dìng yào yǒu CSA rèn zhèng。**
My company is located in California, we only want Prop65 certified products. **我的公司在加州，我们只要Prop65认证过的产品。** **wǒ de gōng sī zài jiā zhōu, wǒ men zhǐ yào Prop65 rèn zhèng guò de chǎn pǐn。**
Can you get an SAA certification for this product? **这个产品可以获得SAA认证吗？** **zhè gè chǎn pǐn kě yǐ huò dé SAA rèn zhèng ma?**
All of the products need to have TRA certification. **所有产品都要有TRA认证。** **suǒ yǒu chǎn pǐn dōu yào yǒu TRA rèn zhèng。**
All of the products passed export inspection, and also have CE and FCC certificates. **所有产品都通过了出口商检，也有CE和FCC认证。** **suǒ yǒu chǎn pǐn dōu tōng guò le chū kǒu shāng jiǎn, yě yǒu CE hé FCC rèn zhèng。**
We are applying for UKCA certification, the process will take six months. **我们正在申请UKCA认证，这个过程要六个月。** **wǒ men zhèng zài shēn qǐng UKCA rèn zhèng, zhè gè guò chéng yào liù gè yuè。**

Quality Inspection - 质检 - **zhì jiǎn**

- Quality Assurance - 质量保证 - **zhì liàng bǎo zhèng**
- Quality Control - 质量控制 - **zhì liàng kòng zhì**
- Major Defect - 主要缺陷 - **zhǔ yào quē xiàn**
- Minor Defect - 轻微缺陷 - **qīng wēi quē xiàn**
- Non - Conformance - 不合格 - **bù hé gé**
- Sample Approval - 样品确认 - **yàng pǐn què rèn**
- Checklist - 检查清单 - **jiǎn chá qīng dān**
- Technical Specification - 技术规格 - **jì shù guī gé**
- Compliance Testing - 合规性测试 - **hé guī xìng cè shì**
- Durability Testing - 耐久性测试 - **nài jiǔ xìng cè shì**
- Safety Testing - 安全测试 - **ān quán cè shì**
- Performance Testing - 性能测试 - **xìng néng cè shì**
- Waterproof Testing - 防水测试 - **fáng shuǐ cè shì**
- Certificate of Conformity (CoC) - 合格证书 - **hé gé zhèng shū**
- Test Report - 测试报告 - **cè shì bào gào**
- Rework - 返工 - **fǎn gōng**
- Workmanship - 工艺质量 - **gōng yì zhì liàng**
- Compliance - 合规性 - **hé guī xìng**

Questions:

How many people are in your QA department? **你们的质量保证部有多少人？** **nǐ men de zhì liàng bǎo zhèng bù yǒu duō shǎo rén?**
What kind of testing do you do in a quality inspection? **你们的质检包括哪些测试？** **nǐ men de zhì jiǎn bāo kuò nǎ xiē cè shì?**
Can you provide testing reports? **你们能提供检验报告吗？** **nǐ men néng tí gōng jiǎn yàn bào gào ma?**
What is the minor defect rate of this product? **这个产品的轻微缺陷率是多少？** **zhè gè chǎn pǐn de qīng wēi quē xiàn lǜ shì duō shǎo?**
Are there any major defects in this batch? **这个批次有没有主要缺陷？** **zhè gè pī cì yǒu méi yǒu zhǔ yào quē xiàn?**
Can I see the inspection checklist? **我可以看看检验清单吗？** **wǒ kě yǐ kàn kàn jiǎn yàn qīng dān ma?**
How do you handle defective items? **你们怎么处理有瑕疵的货？** **nǐ men zěn me chǔ lǐ yǒu xiá cī de huò?**
Do you have ISO certification? **你有没有ISO认证？** **nǐ yǒu méi yǒu ISO rèn zhèng?**
We need tighter quality control. **我们的品控要加强。** **wǒ men de pǐn kòng yào jiā qiáng。**
We need a third party inspection. **我们要第三方检验。** **wǒ men yào dì sān fāng jiǎn yàn。**

Major Defects - 主要缺陷 -**zhǔ yào quē xiàn**

Cracks **裂缝** **liè fèng**	**Delamination** 分层 **fēn céng**	**Inconsistent Quality** 质量不一致 **zhì liàng bù yí zhì**	**Corrosion** **腐蚀** **fǔ shí**
Water Infiltration **渗水** **shèn shuǐ**	**Defective Finish** 表面缺陷 **biǎo miàn quē xiàn**	**Weakness in Load Bearing Elements** 承重部位强度不足 **chéng zhòng bù wèi qiáng dù bù zú**	**Contamination** 污染 **wū rǎn**
Improper Specification 不符合规格 **bù fú hé guī gé**	**Shrinkage** 收缩 **shōu suō**	**Expansion** 膨胀 **péng zhàng**	

Minor Defects - 轻微缺陷 - **qīng wēi quē xiàn**

Surface Irregularities 表面不规则 **biǎo miàn bù guī zé**	**Colour Variation** 颜色变化 **yán sè biàn huà**	**Slight Warping** 轻微翘曲 **qīng wēi qiáo qū**	**Small Cracks** 小裂缝 **xiǎo liè fèng**
Dirt and Contamination 污垢和污染 **wū gòu hé wū rǎn**	**Misalignment** 错位 **cuò wèi**	**Surface Defects** 表面缺陷 **biǎo miàn quē xiàn**	**Minor Chips** 小缺口 **xiǎo quē kǒu**
Inconsistent Texture 纹理不一致 **wén lǐ bù yí zhì**	**Sealant Gaps** 密封剂有间隙 **mì fēng jì yǒu jiàn xì**		

Defective Rate - 缺陷率 - **quē xiàn lǜ**

Defective Rate	Is	Percentage.
The minor defective rate of this batch 这个批次的轻微缺陷率 zhè gè pī cì de qīng wēi quē xiàn lǜ	is 是 shì	5%. 百分之五。 bǎi fēn zhī wǔ。
The major defective rate of this batch 这批货的主要缺陷率 zhè pī huò de zhǔ yào quē xiàn lǜ	is 是 shì	3%. 百分之三。 bǎi fēn zhī sān。
The non-compliance rate of the goods 货物的不合规格率 huò wù de bù hé guī lǜ	is 是 shì	4%. 百分之四。 bǎi fēn zhī sì。
The non-compliance rate of this batch 这个批次的不合格率 zhè gè pī cì de bù hé gé lǜ	is 是 shì	15%. 百分之十五。 bǎi fēn zhī shí wǔ。

Percentage of Goods	Has	Defects.
1% of the A5 screws 百分之一的A5螺钉 bǎi fēn zhī yī de A5 luó dīng	have 有 yǒu	expansion. 膨胀。 péng zhàng。
10% of the aluminum doors 百分之十的铝合金门 bǎi fēn zhī shí de lǚ hé jīn mén	have 有 yǒu	surface defects. 表面缺陷。 biǎo miàn quē xiàn。
Over half of the wood panels 超过一半的木板 chāo guò yī bàn de mù bǎn	have 有 yǒu	colour variation. 颜色变化。 yán sè biàn huà。
30% of the goods 百分之三十的货 bǎi fēn zhī sān shí de huò	have 有 yǒu	small cracks. 小裂缝。 xiǎo liè fèng。

Examples:

The 5% defective rate is too high. This batch needs rework. **百分之五的缺陷率太高了，这批货需要返工。** **bǎi fēn zhī wǔ de quē xiàn lǜ tài gāo le, zhè pī huò xū yào fǎn gōng。**
The 3% minor defect rate is unacceptable. **我们不可以接受百分之三的轻微缺陷率。** **wǒ men bù kě yǐ jiē shòu bǎi fēn zhī sān de qīng wēi quē xiàn lǜ。**
This batch has very inconsistent quality, can you do a reinspection? **这批货的质量非常不一致，可以复检吗？** **zhè pī huò de zhì liàng fēi cháng bù yí zhì, kě yǐ fù jiǎn ma?**
The client wants to see all the test reports. **客户要看所有检验报告。** **kè hù yào kàn suǒ yǒu jiǎn yàn bào gào。**
We cannot accept more than 1% non-compliance rate in the products. **我们不可以接受产品有超过百分之一的不合格率。** **wǒ men bù kě yǐ jiē shòu chǎn pǐn yǒu chāo guò bǎi fēn zhī yī de bù hé gé lǜ。**
Weakness in load bearing is a major defect and we cannot accept the goods. **承重部位强度不足是主要缺陷，我们不可以接受这样的货物。** **chéng zhòng bù wèi qiáng dù bù zú shì zhǔ yào quē xiàn, wǒ men bù kě yǐ jiē shòu zhè yàng de huò wù。**
Sealant gaps are considered a minor defect, only 1% of the products have this defect. **密封剂有间隙算是轻微缺陷，只有百分之一的产品有这个缺陷。** **mì fēng jì yǒu jiàn xì suàn shì qīng wēi quē xiàn, zhǐ yǒu bǎi fēn zhī yī de chǎn pǐn yǒu zhè gè quē xiàn。**
The colour variation in the ceramic tiles is considered a major defect. **瓷砖有颜色变化是主要缺陷。** **cí zhuān yǒu yán sè biàn huà shì zhǔ yào quē xiàn。**
What kind of defects are found in this inspection report? **这个检验报告发现了什么缺陷？** **zhè gè jiǎn yàn bào gào fā xiàn le shén me quē xiàn?**
How many quality inspections have you done during the production? **你们在生产过程中做了多少次质检？** **nǐ men zài shēng chǎn guò chéng zhōng zuò le duō shǎo cì zhì jiǎn?**

- Factory Audit - 验厂 - **yàn chǎng**
- Social Audit - 社会责任审计 - **shè huì zé rèn shěn jì**
- Labour Rights - 劳工权益 - **láo gōng quán yì**
- Environmental Audit - 环境审计 - **huán jìng shěn jì**
- Environmental Standard - 环境标准 - **huán jìng biāo zhǔn**
- Compliance Audit - 合规审计 - **hé guī shěn jì**
- Social Responsibility - 社会责任- **shè huì zé rèn**
- Quality Audit - 质量审计 - **zhì liàng shěn jì**
- Quality Control Procedures - 质量控制流程 - **zhì liàng kòng zhì liú chéng**

Examples:

Our client requires a third party factory audit. **我们的客户要求第三方验厂。** **wǒ men de kè hù yāo qiú dì sān fāng yàn chǎng。**
We would like to do an environmental audit for your warehouse. **我们想要给你的仓库做一个环境审计。** **wǒ men xiǎng yào gěi nǐ de cāng kù zuò yí gè huán jìng shěn jì。**
We are required by our government to do a social audit on all manufacturers. **政府要求我们对所有厂商做社会审计。** **zhèng fǔ yāo qiú wǒ men duì suǒ yǒu chǎng shāng zuò shè huì shěn jì。**
Have you done a quality audit in the past? **你们以前有没有做过质量审计？** **nǐ men yǐ qián yǒu méi yǒu zuò guò zhì liàng shěn jì。**
We are interested in your quality control procedures. **我们对你的质量控制流程很感兴趣。** **wǒ men duì nǐ de zhì liàng kòng zhì liú chéng hěn gǎn xìng qù。**

Topic Discussion

What kind of inspection and certification do you require when you are importing construction materials into your country?

在你的国家进口建材需要哪些检验和认证呢?

zài nǐ de guó jiā jìn kǒu jiàn cái xū yào nǎ xiē jiǎn yàn hé rèn zhèng ne?

Did you see the inspection reports?

你有没有看检验报告?

nǐ yǒu méi yǒu kàn jiǎn yàn bào gào?

Were there any problems or defects with the products?

产品有没有问题或者缺陷?

chǎn pǐn yǒu méi yǒu wèn tí huò zhě quē xiàn?

What is the defective rate?

缺陷率是多少?

quē xiàn lǜ shì duō shǎo?

Have you ever requested a rework?

你有没有要求过返工？

nǐ yǒu méi yǒu yāo qiú guò fǎn gōng ?

Have you ever returned goods?

你有没有退过货？

nǐ yǒu méi yǒu tuì guò huò?

Lesson 4: Customs and Shipping

Vocabulary A:

- Shipping Terms - 运输条款 -**yùnshū tiáokuǎn**
- Bill of Lading - 提单 - **tí dān**
- Airway Bill - 空运提单 - **kōng yùn tí dān**
- Sea Freight - 海运费 - **hǎi yùn fèi**
- Air Freight - 空运费 - **kōng yùn fèi**
- Freight Forwarder - 货运代理 - **huò yùn dài lǐ**
- Container - 集装箱 - **jí zhuāng xiāng**
- Full Container Load (**FCL**) - 整箱货运 - **zhěng xiāng huò yùn**
- Less Than Container Load (LCL) - 拼箱货运 - **pīn xiāng huò yùn**
- Cubic Metre (CBM) - 立方米 - **lì fāng mǐ**
- Container Yard (CY) - 集装箱场地 - **jí zhuāng xiāng chǎng dì**
- Container Freight Station (CFS) - 集装箱货物拼装场 - **jí zhuāng xiāng huò wù pīn zhuāng chǎng**
- Carriage Paid To (CPT) - 运费付至 - **yùn fèi fù zhì**
- Cost and Freight (CFR) - 成本加运费价 - **chéng běn jiā yùn fèi jià**
- Delivered At Place (DAP) - 目的地交货- **mù dì dì jiāo huò**
- Delivered Duty Paid (DDP) - 完税后交货 - **wán shuì hòu jiāo huò**
- Estimated Time of Arrival - 预计到达时间 - **yù jì dào dá shí jiān**

- Estimated Time of Departure - 预计出发时间 - **yù jì chū fā shí jiān**
- Shipping Mark - 唛头 - **mài tóu**
- Freight Collect - 到付运费 - **dào fù yùn fèi**
- Freight Prepaid - 预付运费 - **yù fù yùn fèi**
- Transit Time - 运输时间 - **yùn shū shí jiān**
- Telex Release - 电放 - **diàn fàng**
- Warehousing - 仓储 - **cāng chǔ**
- Less Than Truckload (LTL) - 零担货物 - **líng dān huò wù**

Packaging:

Inner Packing **内包装** **nèi bāo zhuāng**	**Outer Packing** **外包装** **wài bāo zhuāng**	**Carton** **纸箱** **zhǐ xiāng**	**Pallet** **托盘** **tuō pán**
Crate **板条箱** **bǎn tiáo xiāng**	**Wooden Box** **木箱** **mù xiāng**	**Bubble Wrap** **气泡膜** **qì pào mó**	**Foam** **泡沫** **pào mò**
Corrugated Box **瓦楞纸板箱** **wǎ léng zhǐ bǎn xiāng**	**Plastic Bag** **塑料袋** **sù liào dài**	**20GP** **20英尺小柜** **èr shí yīng chǐ xiǎo guì**	**40GP** **40英尺大柜** **sì shí yīng chǐ dà guì**
45GP **45英尺高柜** **sì shí wǔ yīng chǐ gāo guì**	**Tare Weight** **箱体自重** **xiāng tǐ zì zhòng**	**Tare Container** **空箱** **kōng xiāng**	**Loaded Container/Cargo** **重箱** **zhòng xiāng**
DC (Dry Container) **干货柜** **gān huò guì**	**OT (Open Top)** **开顶柜** **kāi dǐng guì**		

Find the right freight forwarder for your shipments:

寻找合适的货代：

xún zhǎo hé shì de huò dài:

Questions to ask the freight forwarders:

1	**How long have you been in the freight forwarding industry?** **你在货代行业有多少年经验了？** **nǐ zài huò dài háng yè yǒu duō shǎo nián jīng yàn le?**
2	**Are you experienced with shipments of construction materials?** **你有没有建材或者相关的货物运输经验？** **nǐ yǒu méi yǒu jiàn cái huò zhě xiāng guān de huò wù yùn shū jīng yàn?**
3	**What services do you offer?** **你能提供哪些服务？** **nǐ néng tí gōng nǎ xiē fú wù?**
4	**What is the typical shipping time for shipment to Canada?** **运输到加拿大一般要多长时间？** **yùn shū dào jiā ná dà yī bān yào duō cháng shí jiān?**
5	**How do you handle delays or problems during transit?** **你如何处理运输途中遇到的意外延误和问题？** **nǐ rú hé chǔ lǐ yùn shū tú zhōng yù dào de yì wài yán wù hé wèn tí?**
6	**Can you provide a detailed breakdown of your price?** **你能提供详细的定价结构吗？** **nǐ néng tí gōng xiáng xì de dìng jià jié gòu ma?**
7	**Are there any additional fees?** **有没有可能会产生的额外的费用？** **yǒu méi yǒu kě néng huì chǎn shēng de é wài de fèi yòng?**
8	**Do you have experience dealing with Canadian customs clearance?** **你有处理加拿大清关的经验吗？** **nǐ yǒu chǔ lǐ jiā ná dà qīng guān de jīng yàn ma?**
9	**Are you familiar with Canadian regulations and compliances?** **你熟悉加拿大的法规和合规要求吗？** **nǐ shúxī Jiānádà de fǎguī hé héguī yāoqiú ma?**

10	**How do I track my shipments?** **我要怎么追踪我的货物？** **wǒ yào zěn me zhuī zōng wǒ de huò wù?**
11	**What insurance options do you have for shipments?** **你能为货物提供哪些保险？** **nǐ néng wèi huò wù tí gōng nǎ xiē bǎo xiǎn?**
12	**Can you accommodate changes in shipment volume or frequency?** **你们能适应运输量和频率的变化吗？** **nǐ men néng shì yìng yùn shū liàng hé yùn shū pín lǜ de biàn huà ma?**
13	**Can you handle peak season and sudden increase in demand?** **你能处理高峰期或者突发的需求增加吗？** **nǐ néng chǔ lǐ gāo fēng qī huò zhě tū fā de xū qiú zēng jiā ma?**
14	**Do you offer fixed rates for long term contracts?** **长期合同有固定价格吗？** **cháng qī hé tóng yǒu gù dìng jià gé ma?**

Do you have a freight forwarder? What kind of services do they provide?

你有货代吗？ 他们提供什么服务？

nǐ yǒu huò dài ma? tā men tí gōng shén me fú wù?

Services	Answer the questions below:
Transportation Arrangement 运输安排 yùn shū ān pái	What kind of transportation do you usually use for shipment? 你的货物一般用哪种运输方式？ nǐ de huò wù yī bān yòng nǎ zhǒng yùn shū fāng shì?
Customs Brokerage 海关代理 hǎi guān dài lǐ	Do you know which documents are needed for customs clearance? 你知道清关需要哪些文件吗？ nǐ zhī dào qīng guān xū yào nǎ xiē wén jiàn ma?
Cargo Insurance 货物保险 huò wù bǎo xiǎn	Which kinds of insurance do you need when shipping from China? 从中国运输的货物需要买哪些保险？ cóng zhōng guó yùn shū de huò wù xū yào mǎi nǎ xiē bǎo xiǎn?
Warehousing and Storage 仓储和储存 cāng chǔ hé chǔ cún	Does the freight forwarder have warehousing services? 你的货代提供仓储服务吗？ nǐ de huò dài tí gōng cāng chǔ fú wù ma?
Consolidation Service 合并服务 hé bìng fú wù / pīn xiāng fú wù	Do you have small shipments that require consolidation service? 你有没有需要拼箱的货物？ nǐ yǒu méi yǒu xū yào pīn xiāng de huò wù?
Documentation Handling 文件处理 wén jiàn chǔ lǐ	Which documents does your freight forwarder prepare for you? 你的货代一般要准备哪些文件？ nǐ de huò dài yī bān yào zhǔn bèi nǎ xiē wén jiàn?
Order Tracking 订单跟踪 dìng dān gēn zōng	Do you track your shipments? 你会跟踪订单吗？ nǐ huì gēn zōng dìng dān ma?
Consultation Services 咨询服务 zī xún fú wù	Do you think your freight forwarder gives valuable advice? 你认为你的货代能提供有价值的信息吗？ nǐ rèn wéi nǐ de huò dài néng tí gōng yǒu jià zhí de xìn xī ma?
Value-Added Services 增值服务 zēng zhí fú wù	Does your freight forwarder provide any value- added services? If so, what are they? 你的货代提供增值服务吗？ 如果有，是哪些呢？

	nǐ de huò dài tí gōng zēng zhí fú wù ma? rú guǒ yǒu, shì nǎ xiē ne?
E-Commerce Logistics **电商物流** **diàn shāng wù liú**	**Does your business require E-Commerce Logistics?** **你的公司有电商物流的需求吗？** **nǐ de gōng sī yǒu diàn shāng wù liú de xū qiú ma?**
Expediting Services **加急服务** **jiā jí fú wù**	**Have you ordered expediting service in the past? Were there additional fees?** **你过去有没有用过加急服务？ 有没有额外的费用？** **nǐ guò qù yǒu méi yǒu yòng guò jiā jí fú wù? yǒu méi yǒu é wài de fèi yòng?**

- Customs Declaration - 报关 - **bào guān**
- Customs Clearance - 清关 - **qīng guān**
- Customs Duty - 关税 - **guān shuì**
- Import Tariffs - 进口关税 - **jìn kǒu guān shuì**
- Export Tariffs - 出口关税 - **chū kǒu guān shuì**
- Contract - 合同 - **hé tóng**
- Invoice - 发票 - **fā piào**
- Packing List - 装箱单 - **zhuāng xiāng dān**
- Bill of Lading - 提单 - **tí dān**

Informations on Bill of Lading , B/L

B/L No. 提单号 tí dān hào	**Carrier** 承运人 chéng yùn rén	**Forwarder** 货代 huò dài	**Shipper** 发货人 fā huò rén
Consignee 收货人 shōu huò rén	**Notify Party** 通知人 tōng zhī rén	**Port of Loading (POL)** 装货港 zhuāng huò gǎng	**Port of Discharge (POD)** 卸货港 xiè huò gǎng
Final Destination 最终目的地 zuì zhōng mù dì dì	**Vessel/Voyage No.** 船名和船次 chuán míng hé chuán cì	**Date of Issue** 提单签发日期 tí dān qiān fā rì qī	**On Board Date/Shipped** 装船日期 zhuāng chuán rì qī
Shipping Marks 标记唛头 biāo jì mài tóu	**Description of Goods** 货物描述 huò wù miáo shù	**No. of Packages** 包装件数 bāo zhuāng jiàn shù	**Gross Weight** 毛重 máo zhòng
Measurement/CBM 体积/立方米 tǐ jī / lì fāng mǐ	**Freight Terms** 运费条款 yùn fèi tiáo kuǎn	**Prepaid** 已付 yǐ fù	**Collect** 到付 dào fù
Freight Amount 运费金额 yùn fèi jīn é	**Additional Charges** 其他费用 qí tā fèi yòng	**Place of Issue** 提单签发地 tí dān qiān fā dì	**Number of Original B/Ls** 提单份数 tí dān fèn shù
Issued By 签发人 qiān fā rén	**Type of B/L** 提单类型 tí dān lèi xíng		

Customs - 海关 - **hǎi guān**

- Customs Broker - 报关行 - **bào guān háng**
- Value Added Tax (VAT) - 增值税 - **zēng zhí shuì**
- Consumption Tax - 消费税 - **xiāo fèi shuì**
- Air Waybill - 空运提单 - **kōng yùn tí dān**
- Certificate of Origin - 原产地证 - **yuán chǎn dì zhèng**
- GSP Certificate of Origin - 普惠制原产地证 - **pǔ huì zhì yuán chǎn dì zhèng**
- Fumigation Certificate - 熏蒸证书 - **xūn zhēng zhèng shū**
- Health Certificate - 卫生证书 - **wèi shēng zhèng shū**
- Certificate of Quality - 品质证书 - **pǐn zhì zhèng shū**
- Certificate of Quantity - 数量证书 - **shù liàng zhèng shū**
- Customs Declaration Form - 报关单 - **bào guān dān**
- Declared Value - 申报价值 - **shēn bào jià zhí**
- Actual Value - 实际价值 - **shí jì jià zhí**
- Dutiable Value - 完税价格 - **wán shuì jià gé**
- Duty - Free - 免税 - **miǎn shuì**
- Duty Reduction - 减税 - **jiǎn shuì**
- Bonded Area - 保税区 - **bǎo shuì qū**
- Bonded Warehouse - 保税仓库 - **bǎo shuì cāng kù**
- Customs Inspection - 海关查验 - **hǎi guān chá yàn**
- Carton Opening Inspection - 开箱查验 - kāi xiāng chá yàn

- Detention - 扣留 - **kòu liú**
- Customs Release - 放行 - **fàng xíng**
- Declaration Delay Fee - 滞报金 - **zhì bào jīn**
- Late Payment Fee - 滞纳金 - **zhì nà jīn**
- Storage Fee - 仓储费 - **cāng chǔ fèi**
- Demurrage - 滞港费 - **zhì gǎng fèi**
- Container Detention Charge - 滞箱费 - **zhì xiāng fèi**
- Commodity Inspection Declaration - 报检 - **bào jiǎn**
- Commodity Inspection - 商检 - **shāng jiǎn**
- Entry - Exit Inspection and Quarantine - 出入境检验检疫 - **chū rù jìng jiǎn yàn jiǎn yì**
- HS code - HS编码 - **HS biān mǎ**
- Commodity code - 商品编码 - **shāng pǐn biān mǎ**
- Import License - 进口许可证 - **jìn kǒu xǔ kě zhèng**
- Export License - 出口许可证 - **chū kǒu xǔ kě zhèng**
- Quota - 配额 - **pèi é**
- Import Prohibition - 禁止进口 - **jìn zhǐ jìn kǒu**
- Import Restriction - 限制进口 - **xiàn zhì jìn kǒu**
- Customs Transit - 转关 - **zhuǎn guān**
- Remote Customs Declaration - 异地报关 - **yì dì bào guān**
- Agent Based Customs Declaration - 代理报关 - **dài lǐ bào guān**

- Self Handled Customs Declaration - 自理报关 - **zì lǐ bào guān**
- Customs Clearance Documents - 清关文件 - **qīng guān wén jiàn**
- Customs Release Notice - 放行通知 - **fàng xíng tōng zhī**

Ask your freight forwarder about customs clearance:

1	**What documents are required for the customs clearance of this shipment?** **这批货物清关需要哪些文件？** **Zhè pī huòwù qīngguān xūyào nǎxiē wénjiàn?**
2	**Do we need a customs broker to handle the customs declaration procedures?** **我们需要委托报关行办理报关手续吗？** **Wǒmen xūyào wěituō bàoguānháng bànlǐ bàoguān shǒuxù ma?**
3	**Does the declared value on this commercial invoice need to match the actual value?** **这份商业发票的申报价值要与实际价值一致吗？** **Zhè fèn shāngyè fāpiào de shēnbào jiàzhí yào yǔ shíjì jiàzhí yízhì ma?**
4	**Will the customs open the cartons to inspect the goods during the customs inspection?** **海关检验时会开箱查验货物吗？** **Hǎiguān jiǎnyàn shí huì kāixiāng cháyàn huòwù ma?**
5	**Will the lack of a certificate of origin affect the customs release of the goods?** **没有原产地证会影响货物放行吗？** **Méiyǒu yuánchǎndì zhèng huì yǐngxiǎng huòwù fàngxíng ma?**
6	**Do we need to pay value-added tax and customs duty for importing this equipment?** **进口这批设备需要缴纳增值税和关税吗？** **Jìnkǒu zhè pī shèbèi xūyào jiǎonà zēngzhíshuì hé guānshuì ma?**
7	**What is the difference in usage between an air waybill and a bill of lading?** **空运提单和海运提单的用途有什么不同？** **Kōngyùn tídān hé hǎiyùn tídān de yòngtú yǒu shénme bùtóng?**
8	**The goods have been detained by the customs: what should we do to handle this?** **货物被海关扣留了，要怎么处理？** **Huòwù bèi hǎiguān kòuliú le, yào zěnme chǔlǐ?**
9	**Can goods in the bonded area be exempted from paying customs duty temporarily?** **保税区的货物可以暂时不缴纳关税吗？** **Bǎoshuìqū de huòwù kěyǐ zànshí bù jiǎonà guānshuì ma?**
10	**Can we handle the self-handled customs declaration or must we find an agent?** **我们可以自己办理自理报关吗？ 还是必须找代理？** **Wǒmen kěyǐ zìjǐ bànlǐ zìlǐ bàoguān ma? Háishì bìxū zhǎo dàilǐ?**
11	**Is the demurrage calculated on a daily basis or based on the weight of the goods?** **滞港费是按天计算还是按货物重量计算的？**

	Zhìgǎngfèi shì àn tiān jìsuàn háishì àn huòwù zhòngliàng jìsuàn de?
12	**Can the GSP Certificate of Origin help us reduce or exempt part of the customs duty?** **普惠制原产地证可以帮我们免部分关税吗？** **Pǔhuìzhì yuánchǎndì zhèng kěyǐ bāng wǒmen miǎn bùfèn guānshuì ma?**
13	**What additional materials need to be prepared in advance for remote customs declaration?** **异地报关需要提前准备哪些额外资料？** **Yìdì bàoguān xūyào tíqián zhǔnbèi nǎxiē éwài zīliào?**
14	**Must goods with wooden packaging provide a fumigation certificate?** **木质包装的货物必须提供熏蒸证书吗？** **Mùzhì bāozhuāng de huòwù bìxū tígōng xūnzhēng zhèngshū ma?**
15	**After the customs release notice is issued, how soon can we pick up the goods?** **放行通知下来后，多久能提到货物？** **Fàngxíng tōngzhī xiàlái hòu, duōjiǔ néng tí dào huòwù?**
16	**If the import quota is used up, can we continue to import this type of commodity?** **进口配额用完了，还能继续进口该类商品吗？** **Jìnkǒu pèi'é yòng wán le, hái néng jìxù jìnkǒu gāi lèi shāngpǐn ma?**
17	**Do we need to submit the health certificate and certificate of quality when declaring for commodity inspection?** **报检时需要提交卫生证书和品质证书吗？** **Bàojiǎn shí xūyào tíjiāo wèishēng zhèngshū hé pǐnzhì zhèngshū ma?**
18	**Will a declaration delay fee be incurred if the declaration deadline is exceeded?** **超过申报期限会产生滞报费用吗？** **Chāoguò shēnbào qīxiàn huì chǎnshēng zhìbào fèiyòng ma?**
19	**Will an incorrect declaration of the HS code lead to failure in the customs inspection?** **HS编码报错会导致海关查验不通过吗？** **HS biānmǎ bàocuò huì dǎozhì hǎiguān cháyàn bù tōngguò ma?**

If you work directly with a customs broker, these are some common questions to ask:

1	**What specific documents do I need to provide for my shipment?** **我要为货物提供哪些文件？** **Wǒ yào wèi huòwù tígōng nǎxiē wénjiàn?**
2	**Are there any additional documents required for regulated goods?** **受监管的商品有其他文件要求吗？** **Shòu jiānguǎn de shāngpǐn yǒu qítā wénjiàn yāoqiú ma?**
3	**What duties and taxes will I need to pay for my shipment?** **我需要为我的货物支付哪些关税和税费？** **Wǒ xūyào wèi wǒ de huòwù zhīfù nǎxiē guānshuì hé shuìfèi?**
4	**How are duties calculated, are there any exemptions and reductions available?** **关税是如何计算的，是否有任何豁免或减免可用？** **Guānshuì shì rúhé jìsuàn de, shìfǒu yǒu rènhé huòmiǎn huò jiǎnmiǎn kěyòng?**
5	**How do I determine the correct HS codes for my products?** **我要怎么为我的产品找到正确HS编码？** **Wǒ yào zěnme wèi wǒ de chǎnpǐn zhǎodào zhèngquè HS biānmǎ?**
6	**Can you assist me in classifying my goods properly?** **您能帮助我正确分类我的货物吗？** **Nín néng bāngzhù wǒ zhèngquè fēnlèi wǒ de huòwù ma?**
7	**Are there any specific regulations or restrictions related to my imported goods?** **与我进口的商品相关的特定法规或限制是什么？** **Yǔ wǒ jìnkǒu de shāngpǐn xiāngguān de tèdìng fǎguī huò xiànzhì shì shénme?**
8	**What are the potential compliance risks I should be aware of?** **我应该注意哪些潜在的合规风险？** **Wǒ yīnggāi zhùyì nǎxiē qiánzài de héguī fēngxiǎn?**
9	**What is the typical timeline for customs clearance for my shipment?** **我们的货物通关一般要多长时间？** **Wǒmen de huòwù tōngguān yībān yào duō cháng shíjiān?**
10	**Are there any potential delays I should anticipate?** **可能有哪些潜在延误？** **Kěnéng yǒu nǎxiē qiánzài yánwù?**
11	**Do I need an import license for my goods?** **我是否需要为我的货物申请进口许可证？**

	Wǒ shìfǒu xūyào wèi wǒ de huòwù shēnqǐng jìnkǒu xǔkězhèng?
12	Can you assist in obtaining any necessary licenses? 你能帮我取得相关的许可证吗？ Nǐ néng bāng wǒ qǔdé xiāngguān de xǔkězhèng ma?
13	Can you provide a detailed estimate of your fees and additional costs I should expect? 你能提供你的费用和其他费用的详细估算吗？ Nǐ néng tígōng nǐ de fèiyòng hé qítā fèiyòng de xiángxì gūsuàn ma?
14	What payment methods do you accept for your services? 有哪些付款方式？ Yǒu nǎxiē fùkuǎn fāngshì?
15	How can I prepare for a potential customs audit? 我应该如何准备海关审计？ Wǒ yīnggāi rúhé zhǔnbèi hǎiguān shěnjì?
16	What records should I keep to ensure compliance? 我应该保留哪些记录以确保合规？ Wǒ yīnggāi bǎoliú nǎxiē jìlù yǐ quèbǎo héguī?
17	How can I track the status of my shipment during customs clearance? 我如何在报关过程中查看我的货物状态？ Wǒ rúhé zài bàoguān guòchéng zhōng chákàn wǒ de huòwù zhuàngtài?
18	Will you provide updates throughout the process? 在整个过程中，您会提供更新信息吗？ Zài zhěnggè guòchéng zhōng, nín huì tígōng gēngxīn xìnxī ma?
19	Do you have any suggestions for future shipments? Are there any areas that need improvement? 您对未来的货物报关清关有哪些建议呢？有没有需要改进的地方？ Nín duì wèilái de huòwù bàoguān qīngguān yǒu nǎxiē jiànyì ne? Yǒu méiyǒu xūyào gǎijìn de dìfāng?

Lesson 5: Building Relationships

Vocabulary A:

- Status - 状态 - **Zhuàngtài**
- Development - 进展 - **Jìnzhǎn**
- Delay - 延迟 - **Yánchí**
- Understand - 理解 - **Lǐjiě**
- Make Sure - 确保 - **Quèbǎo**
- Progress - 进度 - **Jìndù**
- On time - 按时 - **Ànshí**
- Department - 部门 - **Bùmén**
- Communicate - 沟通 - **Gōutōng**
- Expedite - 加快 - **Jiākuài**
- In Order To - 以 - **Yǐ**
- Lower - 降低 - **Jiàngdī**
- Impact - 影响 - **Yǐngxiǎng**
- Confirm - 确认 - **Quèrèn**
- News - 消息 - **Xiāoxī**
- Prompt - 及时 - **Jíshí**

Grammar A:

Make sure - 确保 - **quèbǎo**

Make Sure	(Promise/Goal)
Make sure 确保 què bǎo	the goods are delivered on time. 货物按时送达。 huòwù ànshí sòngdá。
Make sure 确保 què bǎo	the products pass inspection. 产品通过检验。 chǎnpǐn tōngguò jiǎnyàn。
Make sure 确保 què bǎo	the clients are satisfied. 客户满意。 kèhù mǎnyì。

Examples:

You need to make sure the goods are delivered on time. **你要确保货物按时送达。** **Nǐ yào quèbǎo huòwù ànshí sòngdá。**
Can you make sure the products pass inspection? **你可以确保产品通过检验吗？** **Nǐ kěyǐ quèbǎo chǎnpǐn tōngguò jiǎnyàn ma?**
My job is to make sure the clients are happy. **我的工作是确保客户满意。** **Wǒ de gōngzuò shì quèbǎo kèhù mǎnyì。**

Confirm - 确认 - **Quèrèn**

Confirm	Information
Confirm **确认** **Què rèn**	**Dates** **日期** **Rìqī**
Confirm **确认** **Què rèn**	**Prices** **价钱** **Jiàqián**
Confirm **确认** **Què rèn**	**Product Information On the Document** **文件里的产品信息** **Wénjiàn lǐ de chǎnpǐn xìnxī**

Example:

Can you confirm the total cost of this shipment? **你可以确认这批货的总价吗？** **Nǐ kěyǐ quèrèn zhè pī huò de zǒngjià ma?**
I am sorry, shipping dates are not yet confirmed. **对不起，运输日期还未确认。** **Duìbùqǐ, yùnshū rìqī hái wèi quèrèn。**
Please have a look at the document and confirm the product information on it. **请看这份文件并确认文件里的产品信息。** **Qǐng kàn zhè fèn wénjiàn bìng quèrèn wénjiàn lǐ de chǎnpǐn xìnxī。**

Speed Up - 加快 - **Jiākuài**

Speed Up	**(x)**
Speed Up **加快** **Jiākuài**	**The Process** **进度** **Jìndù**
Speed Up **加快** **Jiākuài**	**The Production** **生产** **Shēngchǎn**
Speed Up **加快** **Jiākuài**	**The Customs Declaration Procedure** **报关程序** **Bàoguān chéngxù**

Examples:

This is an urgent order, can you speed up the process? **这是紧急订单，你能加快进度吗？** **Zhè shì jǐnjí dìngdān, nǐ néng jiākuài jìndù ma?**
I am sorry, we cannot speed up the production because it may compromise the quality of the products. **很抱歉，我们无法加快生产，因为这有可能降低产品的质量。** **Hěn bàoqiàn, wǒmen wúfǎ jiākuài shēngchǎn, yīnwèi zhè yǒu kěnéng jiàngdī chǎnpǐn de zhìliàng。**
They can expedite the customs declaration for an additional fee. **加快报关过程要交额外的费用。** **Jiākuài bàoguān guòchéng yào jiāo éwài de fèiyòng。**

Read the dialogue below:

Buyer (Importer)	**Hello Lisa, just checking in on the shipment status, any updates?** **你好丽莎，我想了解一下出货状态，有没有新进展呢？** **Nǐ hǎo Lìshā, wǒ xiǎng liǎojiě yíxià chūhuò zhuàngtài, yǒu méiyǒu xīn jìnzhǎn ne?**
Seller (Manufacturer)	**I am very sorry, there is a problem with quality control on the latest batch, there may be a delay in delivery.** **对不起，最新的一批货在质检方面出了个问题，可能要延迟交货。** **Duìbùqǐ, zuìxīn de yī pī huò zài zhìjiǎn fāngmiàn chūle gè wèntí, kěnéng yào yánchí jiāohuò。**
Buyer (Importer)	**What is the problem?** **有什么问题呢？** **Yǒu shénme wèntí ne?**
Seller (Manufacturer)	**We found a few defects in the pre shipment inspection, this batch will need reworking.** **我们在运输前检验中发现了几个缺陷，这一批货需要返工。** **Wǒmen zài yùnshū qián jiǎnyàn zhōng fāxiànle jǐ gè quēxiàn, zhè yī pī huò xūyào fǎngōng。**
Buyer (Importer)	**I understand that sometimes problems may occur, but please make sure you keep the timeline on track.** **我能理解有时候会出现问题，但是请你确保进度按时完成。** **Wǒ néng lǐjiě yǒu shíhòu huì chūxiàn wèntí, dànshì qǐng nǐ quèbǎo jìndù ànshí wánchéng。**
Seller (Manufacturer)	**I apologize again, I will communicate with our production team to expedite the process.** **我很抱歉，我会跟我们的生产部门沟通并加快过程。** **Wǒ hěn bàoqiàn, wǒ huì gēn wǒmen de shēngchǎn bùmén gōutōng bìng jiākuài guòchéng。**
Buyer (Importer)	**Would it be possible to expedite the next batch to minimize the impact?** **可不可以加快下一批货以降低影响？** **Kě bù kěyǐ jiākuài xià yī pī huò yǐ jiàngdī yǐngxiǎng?**
Seller (Manufacturer)	**Certainly, I will communicate with our production manager on this.** **当然可以，我会跟我们的生产经理沟通此事。** **Dāngrán kěyǐ, wǒ huì gēn wǒmen de shēngchǎn jīnglǐ gōutōng cǐ shì。**
Buyer	**Can you share a revised delivery plan with new dates?**

(Importer)	你能给我发一个新的交货计划和新的日期吗？ Nǐ néng gěi wǒ fā yī gè xīn de jiāohuò jìhuà hé xīn de rìqī ma?
Seller (Manufacturer)	Yes, I will let you know as soon as I confirm the delivery dates. 可以，我确认好新的日期就马上通知你。 Kěyǐ, wǒ quèrèn hǎo xīn de rìqī jiù mǎshàng tōngzhī nǐ。
Buyer (Importer)	If you need any help from my end, please let me know. 如果你需要我这边给你任何帮助，请直接告诉我。 Rúguǒ nǐ xūyào wǒ zhè biān gěi nǐ rènhé bāngzhù, qǐng zhíjiē gàosù wǒ。
Seller (Manufacturer)	Thank you, I will let you know if there's any new updates. 谢谢你，如果有任何新的消息我会马上告诉你。 Xièxiè nǐ, rúguǒ yǒu rènhé xīn de xiāoxī wǒ huì mǎshàng gàosù nǐ。
Buyer (Importer)	I appreciate your prompt and clear communication, talk soon. 我很感谢你能及时沟通，下次再聊。 Wǒ hěn gǎnxiè nǐ néng jíshí gōutōng, xià cì zài liáo。

Vocabulary B:

- Receive (impact) - 受到 - **Shòudào**
- Business - 生意 - **Shēngyì**
- As Much As Possible - 尽量 - **Jǐnliàng**
- Think Of Ways To - 想办法 - **Xiǎng bànfǎ**
- Purchase Order - 采购订单 - **Cǎigòu dìngdān**
- Share - 分享 - **Fēnxiǎng**
- Usually - 一般来说 - **Yībān láishuō**
- Impact - 冲击 - **Chōngjī**
- Relationship - 关系 - **Guānxì**
- Peak Season - 高峰期 - **Gāofēngqī**
- Precious - 宝贵 - **Bǎoguì**
- Grateful - 感激 - **Gǎnjī**
- Consider - 考虑 - **Kǎolǜ**
- Perhaps - 也许 -**Yěxǔ**
- Method - 方式 - **Fāngshì**
- Adjust - 调整 - **Tiáozhěng**
- Classification - 分类 - **Fēnlèi**
- Idea - 想法 - **Xiǎngfǎ**

Grammar B:

be affected by... - 受到. . . . . . 影响 - **shòu dào ......yǐng xiǎng**

Receive	X's	Impact
Receive 受到 shòu dào	New Tariff's 新关税的 xīn guān shuì de	Impact 影响 yǐng xiǎng
Receive 受到 shòu dào	Bad Weather's 恶劣天气的 è liè tiān qì de	Impact 影响 yǐng xiǎng
Receive 受到 shòu dào	The Manufacturer's 厂家的 chǎng jiā de	Impact 影响 yǐng xiǎng

Examples:

Was your business impacted by the new tariff? 你的生意有没有受到关税的影响？ nǐ de shēng yì yǒu méi yǒu shòu dào guān shuì de yǐng xiǎng？
The shipment was delayed due to the impact from bad weather. 这批货物受到恶劣天气的影响而延迟了。 zhè pī huò wù shòu dào è liè tiān qì de yǐng xiǎng ér yán chí le。
Our clients' businesses were impacted by the manufacturers. 我们的客户的生意受到了厂家的影响。 wǒ men de kè hù de shēng yì shòu dào le chǎng jiā de yǐng xiǎng。

Influence - 影响 - **yǐng xiǎng**

A	Toward	B		Influence
Weather 天气 tiān qì	Toward 对 duì	Transportation 运输 yùn shū	的 de	Influence 影响 yǐng xiǎng

Tariff **关税** **guān shuì**	**Toward** **对** **duì**	**Business** **企业** **qǐ yè**	**的** **de**	**Influence** **影响** **yǐng xiǎng**
Inspection Report **检验报告** **jiǎn yàn bào gào**	**Toward** **对** **duì**	**Product** **产品** **chǎn pǐn**	**的** **de**	**Influence** **影响** **yǐng xiǎng**

Examples:

The recent bad weather created a big impact on our shipment. **最近的恶劣天气对我们的货物运输有很大的影响。** **zuì jìn de è liè tiān qì duì wǒ men de huò wù yùn shū yǒu hěn dà de yǐng xiǎng。**
What is the influence the new US tariffs have on businesses? **美国的新关税对企业有什么影响？** **měi guó de xīn guān shuì duì qǐ yè yǒu shén me yǐng xiǎng。**
The inspection report has a very important influence on your products. **这个检验报告对你的产品有很重要的影响。** **zhè gè jiǎn yàn bào gào duì nǐ de chǎn pǐn yǒu hěn zhòng yào de yǐng xiǎng。**

As Much As Possible - 尽量 - **jǐn liàng**

As Much As Possible	**Verb Object**
As Much As Possible **尽量** **jǐn liàng**	**Minimize The Impact** **减低影响** **jiǎn dī yǐng xiǎng**
As Much As Possible **尽量** **jǐn liàng**	**Arrive On Time** **按时到达** **àn shí dào dá**
As Much As Possible **尽量** **jǐn liàng**	**Cut Cost** **减少成本** **jiǎn shǎo chéng běn**

Examples:

We try to minimize the impact as much as possible. **我们尽量减少影响。** **wǒ men jǐn liàng jiǎn shǎo yǐng xiǎng。**
The shipper will do their best to deliver on time. **运输公司尽量按时送达。** **yùn shū gōng sī jǐn liàng àn shí sòng dá。**
Businesses are trying to cut as much cost as possible. **企业尽量减少成本。** **qǐ yè jǐn liàng jiǎn shǎo chéng běn。**

Read the dialogue below:

Importer	**Hi Mary, how is business lately? Have you been impacted by the new tariffs?** 马丽，最近生意怎么样？ 你们有没有受到新关税的影响？ **Mǎ Lì, zuì jìn shēng yì zěn me yàng? nǐ men yǒu méi yǒu shòu dào xīn guān shuì de yǐng xiǎng?**
Exporter	**Yes, we are doing the best we can to minimize the impact.** 有，我们正在想办法尽量减少影响。 **Yǒu, wǒ men zhèng zài xiǎng bàn fǎ jǐn liàng jiǎn shǎo yǐng xiǎng。**
Importer	**The new tariff will impact my PO, too. The 20% tariff increase will really impact our customers, is it possible for the supplier to share the tariff cost with us?** 新关税也会影响我的采购订单。百分之二十的加征关税对我的客户影响真的很大。我想问一下供应商有没有可能与我们分担关税呢？ **Xīn guān shuì yě huì yǐng xiǎng wǒ de cǎi gòu dìng dān. bǎi fēn zhī èr shí de jiā zhēng guān shuì duì wǒ de kè hù yǐng xiǎng zhēn de hěn dà. wǒ xiǎng wèn yí xià gōng yìng shāng yǒu méi yǒu kě néng yǔ wǒ men fēn dān guān shuì ne?**
Exporter	**Traditionally the buyer usually bears the cost of the tariff, but this new tariff policy has a great impact on everyone. You are a valued customer and we have had a very good relationship, I will see what I can do.** 一般来说买家要负责关税的费用，但是这个新的关税对我们所有人都有很大的冲击。你是我们的宝贵客户，我们的关系一直很好，我会尽量帮你的。 **Yì bān lái shuō mǎi jiā yào fù zé guān shuì de fèi yòng, dàn shì zhè gè xīn de guān shuì duì wǒ men suǒ yǒu rén dōu yǒu hěn dà de chōng jī. nǐ shì wǒ men de bǎo guì kè hù, wǒ men de guān xì yì zhí hěn hǎo, wǒ huì jǐn liàng bāng nǐ de。**
Importer	**Thank you, I appreciate that.** 谢谢，我很感激。 **Xiè xiè, wǒ hěn gǎn jī。**
Exporter	**May I suggest alternative shipping options to help lower the cost?** 你们或许可以考虑其他运输方式以降低成本？ **Nǐ men huò xǔ kě yǐ kǎo lǜ qí tā yùn shū fāng shì yǐ jiàng dī chéng běn?**
Importer	**I am already working with my freight forwarder to adjust the shipping cost. Unfortunately this is a peak season for shipping and there aren't many alternatives.** 我已经在和我的货代沟通调整运输费用了，可惜现在是货运的高峰期，我们没有很多选择。 **Wǒ yǐ jīng zài hé wǒ de huò dài gōu tōng tiáo zhěng yùn shū fèi yòng le, kě xī xiàn zài shì huò yùn de gāo fēng qī, wǒ men méi yǒu hěn duō xuǎn zé。**

Exporter	**Have you considered updating the HS codes or classification to reduce the duty impact?** **那你有没有考虑过更新HS编码或调整产品分类以降低关税影响呢？** **Nà nǐ yǒu méi yǒu kǎo lǜ guò gēng xīn HS biān mǎ huò tiáo zhěng chǎn pǐn fēn lèi yǐ jiàng dī guān shuì yǐng xiǎng ne?**
Importer	**That's a good idea, I haven't thought of that.** **我还没想到这个呢，这个想法很好。** **Wǒ hái méi xiǎng dào zhè gè ne, zhè gè xiǎng fǎ hěn hǎo。**
Exporter	**Tomorrow we can meet with a customs broker together to see if they have any suggestions.** **我们明天可以一起去见一个报关行，看看他们怎么说。** **Wǒ men míng tiān kě yǐ yì qǐ qù jiàn yí gè bào guān háng, kàn kan tā men zěn me shuō。**
Importer	**I appreciate your help. See you tomorrow.** **我很感激你的帮助，明天见。** **Wǒ hěn gǎn jī nǐ de bāng zhù, míng tiān jiàn。**

Vocabulary C:

- Check - 查看 -**chá kàn**
- Late - **延迟 - yán chí**
- Contact - 联系 - **lián xì**
- Procedure - 过程 - **guò chéng**
- Detain - 扣留 - **kòu liú**
- Rest - 余下 - **yú xià**
- Rest - 其余 - **qí yú**
- Closely - 密切 - **mì qiè**
- Follow - 追踪 - **zhuī zōng**
- Release - 放行 - **fàng xíng**
- Inform - 通知 - **tōng zhī**
- Control - 控制 - **kòng zhì**

Grammar C:

The Remaining - 余下的 - **yú xià de**

Remaining	**Item**
The Remaining **余下的** **yú xià de**	**Goods** **货物** **huò wù**
The Remaining **余下的** **yú xià de**	**Staff** **员工** **yuán gōng**
The Remaining **余下的** **yú xià de**	**Balance Owing** **欠款** **qiàn kuǎn**

Example:

The remaining goods were detained at the customs. **余下的货物被海关扣留了。** **Yú xià de huò wù bèi hǎi guān kòu liú le。**
The remaining staff were on holiday. **余下的员工放假了。** **Yú xià de yuán gōng fàng jià le。**
I will pay the remaining balance tomorrow. **明天我会付余下的欠款。** **Míng tiān wǒ huì fù yú xià de qiàn kuǎn。**

Read the dialogue below:

Client	**Hi Lisa, I am checking with my shipment. Why is it late?** **你好丽萨，我想查看一下我的货物，怎么延迟了？** **Nǐ hǎo Lì Sà, wǒ xiǎng chá kàn yī xià wǒ de huò wù, zěn me yán chí le?**
Freight Forwarder	**Hello Marco, I just checked with the manufacturer and the shipper. The shipment was delayed due to customs clearance. We can send a portion of the shipment via express, and the rest will follow next week.** **你好马克。我刚刚联系了我们的厂家和运输公司，这批货物目前在清关过程中被扣留了。我们可以先快递一部分的货物，余下的下周会送到。** **Nǐ hǎo Mǎ Kè。 Wǒ gāng gāng lián xì le wǒ men de chǎng jiā hé yùn shū gōng sī, zhè pī huò wù mù qián zài qīng guān guò chéng zhōng bèi kòu liú le. Wǒ men kě yǐ xiān kuài dì yī bù fèn de huò wù, yú xià de xià zhōu huì sòng dào。**
Client	**Okay, when will the express delivery arrive?** **好吧，快递什么时候能到？** **Hǎo ba, kuài dì shén me shí hòu néng dào?**
Freight Forwarder	**The express delivery will arrive in two days.** **快递后天到达。** **Kuài dì hòu tiān dào dá。**
Client	**What about the rest of the shipment?** **其余的货物呢？** **Qí yú de huò wù ne?**
Freight Forwarder	**I will track the rest of the shipment closely and update you as soon as it is released.** **我会密切追踪这批货物，海关一放行我就通知你。** **Wǒ huì mì qiè zhuī zōng zhè pī huò wù, hǎi guān yī fàng xíng wǒ jiù tōng zhī nǐ。**
Client	**Please keep the timeline on track. This shipment needs to arrive before the 15th of this month.** **请按时完成，这批货必须在十五号之前送到。** **Qǐng àn shí wán chéng, zhè pī huò bì xū zài shí wǔ hào zhī qián sòng dào。**
Freight Forwarder	**I apologize for the inconvenience, but when the shipment is in customs it is out of our control.** **很抱歉，但是海关扣留的货物我们控制不了。** **Hěn bào qiàn, dàn shì hǎi guān kòu liú de huò wù wǒ men kòng zhì bù liǎo。**
Client	**I understand, you've been very helpful. Please let me know if customs require additional documents.**

	我明白，你帮了我很大的忙。如果海关需要其他文件请通知我。 **Wǒ míng bái, nǐ bāng le wǒ hěn dà de máng. Rú guǒ hǎi guān xū yào qí tā wén jiàn qǐng tōng zhī wǒ。**
Freight Forwarder	**Thank you, I will. Talk to you soon!** **谢谢，我会的。再聊。** **Xiè xiè, wǒ huì de。 Zài liáo。**

Vocabulary D:

- Receive (Goods) - 收到 -**shōu dào**
- Share - 分享 - **fēn xiǎng**
- Immediately - 马上 - **mǎ shàng**
- Photos - 照片 - **zhào piàn**
- One of The - 其中一个 - **qí zhōng yī gè**
- Crack - 裂痕 - **liè hén**
- Broken - **破损** - **pò sǔn**
- Attached - 附上 - **fù shàng**
- Images - 图片 - **tú piàn**
- Quantity - 数量 - **shù liàng**
- Replacement - 替换 - **tì huàn**
- Send Additional - 补发 - **bǔ fā**
- As Soon As Possible - 尽快 - **jǐn kuài**
- Start - 启动 - **qǐ dòng**
- Air Freight - 空运 - **kōng yùn**
- Additional - 额外 - **é wài**
- React - 应对 - **yìng duì**
- Cause - 引起 - **yǐn qǐ**
- Inconvenience - 不便 - **bù biàn**
- Replace - 替换 - **tì hùan**

Grammar D:

One of the - 其中一个 - **qí zhōng yī gè**

One of the	(x)
One of the 其中一艘 qí zhōng yī sōu	Ships 船 chuán
One of the 其中一个 qí zhōng yī gè	Cartons 箱子 xiāng zi
One of the 其中一个 qí zhōng yī gè	Suppliers 供应商 gōng yìng shāng

Examples:

One of the ships had an accident, half the goods were lost. 其中一艘船发生了意外，丢了一半的货物。 Qí zhōng yī sōu chuán fā shēng le yì wài, diū le yī bàn de huò wù。
One of the cartons had damaged goods. 其中一个箱子里面的货物有损坏。 Qí zhōng yī gè xiāng zi lǐ miàn de huò wù yǒu sǔn huài。
One of the suppliers went out of business last year. 去年其中一个供应商倒闭了。 Qù nián qí zhōng yī gè gōng yìng shāng dǎo bì le。

Replace - 替换 - **tì huàn**

Replace	Items
Replace 替换 tì huàn	Half The Goods 一半的货物 yí bàn de huò wù
Replace 替换	A Supplier 一个供应商

tì huàn	**yí gè gōng yìng shāng**
Replace **替换** **tì huàn**	**Shipper** **运输公司** **yùn shū gōng sī**

Examples:

The client wants to replace half the goods. **客户要替换一半货物。** **Kè hù yào tì huàn yī bàn huò wù。**
We need to replace a supplier. **我们要替换供应商。** **Wǒ men yào tì huàn gōng yìng shāng。**
Is it possible to replace a shipper? **可以替换运输公司吗？** **Kě yǐ tì huàn yùn shū gōng sī ma?**

On Time - 按时（准时） - **àn shí (zhǔn shí)**

On Time	Verb - Object
On Time **按时** **jí shí**	**Got Customs Clearance** **清关** **qīng guān**
On Time **按时** **jí shí**	**Contact You** **联系你** **lián xì nǐ**
On Time **按时** **jí shí**	**Finish Inspection** **完成检验** **wán chéng jiǎn yàn**

Examples:

Your shipment received customs clearance on time! **你的货物按时清关了！** **Nǐ de huò wù jí shí qīng guān le!**
If there's any news, I will contact you right away. **如果有任何消息，我会立刻联系你。** **Rú guǒ yǒu rèn hé xiāo xi, wǒ huì lì kè lián xì nǐ。**
I am very pleased that you finished the inspection on time. **我很高兴你按时完成检验。** **Wǒ hěn gāo xìng nǐ àn shí wán chéng jiǎn yàn。**

Read the dialogue below:

Client	**Lisa, I just received the shipment today, we found some damaged products in the shipment.** **丽莎，我今天刚刚收到货物，这批货里有一些损坏的产品。** **Lì Shā, wǒ jīn tiān gāng gāng shōu dào huò wù, zhè pī huò lǐ yǒu yì xiē sǔn huài de chǎn pǐn。**
Customer Service	**I am sorry to hear about the damage. Could you share photos of the packaging and the damaged items? I will speak with our shipper immediately.** **我很抱歉您的货物有损坏。你可以分享损坏的货物和箱子的照片吗？我马上联系运输公司。** **Wǒ hěn bào qiàn nín de huò wù yǒu sǔn huài. Nǐ kě yǐ fēn xiǎng sǔn huài de huò wù hé xiāng zi de zhào piàn ma? Wǒ mǎ shàng lián xì yùn shū gōng sī。**
Client	**The ceramic tiles in one carton had cracks, some were broken into pieces. Attached are the images.** **其中一个箱子里的瓷砖有裂痕，有些甚至碎成片了。我附上了几张图片。** **Qí zhōng yī gè xiāng zi lǐ de cí zhuān yǒu liè hén, yǒu xiē shèn zhì suì chéng piàn le. Wǒ fù shàng le jǐ zhāng tú piàn。**
Customer Service	**Thank you. I can arrange a replacement order right away or send it together with your next shipment. Can you give me the quantity of the damaged items?** **谢谢。我可以马上给你补发或者放在下一批货里一起发。你能给我损坏的数量吗？** **Xiè xiè. Wǒ kě yǐ mǎ shàng gěi nǐ bǔ fā huò zhě fàng zài xià yī pī huò lǐ yī qǐ fā. Nǐ néng gěi wǒ sǔn huài de shù liàng ma?**
Client	**I need 180 pieces of the A085 tiles replaced, please send them as soon as possible.** **我需要替换一百八十块A085号的瓷砖，请尽快发货。** **Wǒ xū yào tì huàn yī bǎi bā shí kuài A085 hào de cí zhuān, qǐng jǐn kuài fā huò。**
Customer Service	**I can send them out right away and it will arrive next month.** **我可以马上发货，下个月到。** **Wǒ kě yǐ mǎ shàng fā huò, xià gè yuè dào。**
Client	**Can you expedite it? I need them in two weeks as my project is starting next week.** **你可以加快吗？ 我的项目下周启动，我两周后就要用到这些瓷砖。** **Nǐ kě yǐ jiā kuài ma? Wǒ de xiàng mù xià zhōu qǐ dòng, wǒ liǎng zhōu hòu jiù yào yòng dào zhè xiē cí zhuān。**
Customer Service	**Sure, we will ship it by air and it will arrive in one week.** **好的，我们可以空运，下周就到。** **Hǎo de, wǒ men kě yǐ kōng yùn, xià zhōu jiù dào。**

Client	**That is great, I really appreciate your effort. Do you need anything else from me?** **太好了，我很感谢你的帮助。你还需要我这边的东西吗？** **Tài hǎo le, wǒ hěn gǎn xiè nǐ de bāng zhù. Nǐ hái xū yào wǒ zhè biān de dōng xi ma?**
Customer Service	**We may need additional documents for customs declarations, I will let you know if more documents are required.** **报关方面我们可能需要额外的文件，如果需要的话我会尽快通知你。** **Bào guān fāng miàn wǒ men kě néng xū yào é wài de wén jiàn, rú guǒ xū yào de huà wǒ huì jǐn kuài tōng zhī nǐ,**
Client	**Great, I appreciate your immediate response and effort to solve this problem for us,** **好的，我感谢你能及时回应并且解决问题。** **Hǎo de, wǒ gǎn xiè nǐ néng jí shí huí yìng bìng qiě jiě jué wèn tí。**
Customer Service	**You are welcome, I am sorry about the inconvenience that this has caused and I will update you as soon as the replacement tiles are shipped.** **没关系，我很抱歉这批货物的损坏引起了不便。替换的瓷砖发货后我会及时通知你。** **Méi guān xì, wǒ hěn bào qiàn zhè pī huò wù de sǔn huài yǐn qǐ le bù biàn. Tì huàn de cí zhuān fā huò hòu wǒ huì jí shí tōng zhī nǐ。**

What are some common problems that you've met when importing from China? How do you communicate and solve problems?

1. You found defects in the shipment, you say:

2. The shipment is delayed for two weeks. You say:

3. Your goods are held at the customs due to mistakes in documentation, you say:

4. The manufacturer cannot ship the goods on time, you say:

5. You found additional fees on the shipping invoice, you say:

Final Project: Your Purchase Order

Please talk about your recent purchase and answer the following questions:

Which items are on your purchase order?

你的采购订单有哪些产品？

Nǐ de cǎi gòu dìng dān yǒu nǎ xiē chǎn pǐn?

How did you find your supplier?

你怎么找到供应商的？

Nǐ zěn me zhǎo dào gōng yìng shāng de?

Who is your manufacturer? Where are they located?

你的厂家是谁？他们在哪儿？

Nǐ de chǎng jiā shì shéi? Tā men zài nǎ ér?

What is the lead time?

交货时间是多久？

Jiāo huò shí jiān shì duō jiǔ?

Were there any problems with the manufacturing?

生产过程有没有问题？

Shēng chǎn guò chéng yǒu méi yǒu wèn tí?

Do they have all the required certifications?

他们拥有必要的认证吗？

Tā men yōng yǒu bì yào de rèn zhèng ma?

Did the products pass inspections?

产品有没有通过检验？

Chǎn pǐn yǒu méi yǒu tōng guò jiǎn yàn?

What kind of inspections were done?

产品做了哪些检验？

Chǎn pǐn zuò le nǎ xiē jiǎn yàn?

Did you negotiate the price with the suppliers?

你有没有和供应商讲价？

Nǐ yǒu méi yǒu hé gōng yìng shāng jiǎng jià?

Were there any problems with customs clearance?

清关有没有遇到问题？

Qīng guān yǒu méi yǒu yù dào wèn tí?

Did you hire a customs broker?

你有没有请报关行？

Nǐ yǒu méi yǒu qǐng bào guān háng?

Did you work with a freight forwarder? What kind of services do they provide?

你有没有货代？他们提供什么服务？

Nǐ yǒu méi yǒu huò dài? Tā men tí gōng shén me fú wù?

Are you pleased with the freight forwarder's service?

你对货代的服务满意吗？

Nǐ duì huò dài de fú wù mǎn yì ma?

Did the shipment arrive on time?

货物有没有按时到达？

Huò wù yǒu méi yǒu àn shí dào dá?

Are you satisfied with this shipment?

你对这批货物满意吗？

Nǐ duì zhè pī huò wù mǎn yì ma?

Vocabulary List

1 - Supplier - 供应商 - gōng yìng shāNg
2 Manufacturer - 厂商 - chǎng shāng
3 Original Manufacturer - 源头厂商 - yuán tóu chǎng shāng
4 Wholesaler - 批发商 - pī fā shāng
5 Factory - 工厂 - gōng chǎng
6 Warehouse - 仓库 - cāng kù
7 Exporter - 出口代理 - chū kǒu dài lǐ
8 Construction Materials - 建筑材料 - jiàn zhù cái liào
9 Size - 尺寸 - chǐ cùn
10 Length - 长 - cháng
11 Width - 宽 - kuān
12 Depth - 深 - shēn
13 Thickness - 厚度 - hòu
14 Diameter - 直径 - zhí jìng
15 Metre - 米 - mǐ
16 Centimetre - 厘米 - lí mǐ
17 Millimeter -毫米 - háo mǐ
18 KG - 公斤 - gōng jīn
19 Ton - 吨 - dūn
20 Weight - 重量 - zhòng liàng
21 Load Capacity - 承重 - chéng zhòng
22 Type - 型号 - xíng hào
23 Materials - 材料- cái liào
24 Quote - 报价 - bào jià
25 Exchange Rate - 汇率 - huì lǜ
26 MOQ - 最少起订量 - zuì shǎo qǐ dìng liàng
27 Ready Stock - 现货 - xiàn huò
28 Free On Board - 船上交货价 - chuán shàng jiāo huò jià
29 C&F - 成本加运费 - chéng běn jiā yùn fèi
30 CIF (Cost Insurance Freight) - 成本加保险加运费 - chéng běn jiā bǎo xiǎn jiā yùn fèi

31 DDP - 完税后交货 - wán shuì hòu jiāo huò
32 EXW - 出厂价 - chū chǎng jià
33 Purchase Order - 采购订单 - cǎi gòu dìng dān
34 Proforma Invoice - 形式发票 - xíng shì fā piào
35 Commercial Invoice - 商业发票 - shāng yè fā piào
36 Contract - 合同 - hé tóng
37 Letter of Credit (L/C) - 信用证 - xìn yòng zhèng
38 Telegraphic Transfer (T/T) - 电汇 - diàn huì
39 Documents Against Payment (D/P) - 付款交单 - fù kuǎn jiāo dān
40 Payment Method - 付款方式 - fù kuǎn fāng shì
41 Deposit - 订金 - dìng jīn
42 Unit Price - 单价 - dān jià
43 Quantity - 数量 - shù liàng
44 Total - 总价 - zǒng jià
45 Shipping - 运输 - yùn shū
46 Insurance - 保险 - bǎo xiǎn
47 Shipment Value - 货物价值 - huò wù jià zhí
48 Customs and Duties - 关税 - guān shuì
49 Quality Inspection Fee - 检验费 - jiǎn yàn fèi
50 Proforma Invoice - 形式发票 - xíng shì fā piào
51 Commercial Invoice - 商业发票 - shāng yè fā piào
52 Packing List - 装箱单 - zhuāng xiāng dān
53 Discount - 折扣 - zhé kòu
54 Fee - 费用 - fèi yòng
55 Detailed - 详细的 - xiáng xì de
56 Lead Time - 交货时间 - jiāo huò shí jiān
57 Samples - 样品 - yàng pǐn
58 Quality Control - 质量控制 - zhì liàng kòng zhì
59 Production - 生产 - shēng chǎn
60 Return Policy - 退货政策 - tuì huò zhèng cè
61 Return(goods) - 退货 - tuì huò
62 Standard - 标准 - biāo zhǔn
63 Meet Standard - 符合标准 - fú hé biāo zhǔn

64 **Large Orders - 高量订单 - gāo liàng dìng dān**
65 **Flexible - 灵活 - líng huó**
66 **Change - 改变 - gǎi biàn**
67 **Product Specs - 产品规格 - chǎn pǐn guī gé**
68 **Catalog - 产品目录 - chǎn pǐn mù lù**
69 **Competitor - 竞争者 - jìng zhēng zhě**
70 **Defective - 瑕疵 - xiá cī**
71 **Warranty - 保修期 - bǎo xiū qī**
72 **Damaged - 损坏 - sǔn huài**
73 **Inspection Reports - 检验报告 - jiǎn yàn bào gào**
74 **Inspection - 检验 - jiǎn yàn**
75 **Pre-Shipment Inspection - 装运前检验 - zhuāng yùn qián jiǎn yàn**
76 **Export Inspection - 出口商检 - chū kǒu shāng jiǎn**
77 **Quality Inspection - 质检 - zhì jiǎn**
78 **Third Party Inspection 第三方检验 - dì sān fāng jiǎn yàn**
79 **Pre-Production Inspection - 产前检验 - chǎn qián jiǎn yàn**
80 **During-Production Inspection - 生产中检验 - shēng chǎn zhōng jiǎn yàn**
81 **Final Inspection - 最终检验 - zuì zhōng jiǎn yàn**
82 **Container Loading Inspection - 装柜监装 - zhuāng guì jiān zhuāng**
83 **Random Inspection - 随机抽检 - suí jī chōu jiǎn**
84 **Re-inspection - 复检 - fù jiǎn**
85 **Product Certification - 产品认证 - chǎn pǐn rèn zhèng**
86 **Quality Assurance - 质量保证 - zhì liàng bǎo zhèng**
87 **Quality Control - 质量控制 - zhì liàng kòng zhì**
88 **Major Defect - 主要缺陷 - zhǔ yào quē xiàn**
89 **Minor Defect - 轻微缺陷 - qīng wēi quē xiàn**
90 **Non - Conformance - 不合格 - bù hé gé**
91 **Sample Approval - 样品确认 - yàng pǐn què rèn**
92 **Checklist - 检查清单 - jiǎn chá qīng dān**
93 **Technical Specification - 技术规格 - jì shù guī gé**
94 **Compliance Testing - 合规性测试 - hé guī xìng cè shì**
95 **Durability Testing - 耐久性测试 - nài jiǔ xìng cè shì**
96 **Safety Testing - 安全测试 - ān quán cè shì**
97 **Performance Testing - 性能测试 - xìng néng cè shì**

98 **Waterproof Testing - 防水测试 - fáng shuǐ cè shì**
99 **Certificate of Conformity (CoC) - 合格证书 - hé gé zhèng shū**
100 **Test Report - 测试报告 - cè shì bào gào**
101 **Rework - 返工 - fǎn gōng**
102 **Workmanship - 工艺质量 - gōng yì zhì liàng**
103 **Compliance - 合规性 - hé guī xìng**
104 **Factory Audit - 验厂 - yàn chǎng**
105 **Social Audit - 社会责任审计 - shè huì zé rèn shěn jì**
106 **Labour Rights - 劳工权益 - láo gōng quán yì**
107 **Environmental Audit - 环境审计 - huán jìng shěn jì**
108 **Environmental Standard - 环境标准 - huán jìng biāo zhǔn**
109 **Compliance Audit - 合规审计 - hé guī shěn jì**
110 **Social Responsibility - 社会责任- shè huì zé rèn**
111 **Quality Audit - 质量审计 - zhì liàng shěn jì**
112 **Quality Control Procedures - 质量控制流程 - zhì liàng kòng zhì liú chéng**
113 **Shipping Terms - 运输条款 -yùnshū tiáokuǎn**
114 **Bill of Lading - 提单 - tí dān**
115 **Airway Bill - 空运提单 - kōng yùn tí dān**
116 **Sea Freight - 海运费 - hǎi yùn fèi**
117 **Air Freight - 空运费 - kōng yùn fèi**
118 **Freight Forwarder - 货运代理 - huò yùn dài lǐ**
119 **Container - 集装箱 - jí zhuāng xiāng**
120 **Full Container Load (FCL) - 整箱货运 - zhěng xiāng huò yùn**
121 **Less Than Container Load (LCL) - 拼箱货运 - pīn xiāng huò yùn**
122 **Cubic Metre (CBM) - 立方米 - lì fāng mǐ**
123 **Container Yard (CY) - 集装箱场地 - jí zhuāng xiāng chǎng dì**
124 **Container Freight Station (CFS) - 集装箱货物拼装场 - jí zhuāng xiāng huò wù pīn zhuāng chǎng**
125 **Carriage Paid To (CPT) - 运费付至 -yùn fèi fù zhì**
126 **Cost and Freight (CFR) - 成本加运费价 - chéng běn jiā yùn fèi jià**
127 **Delivered At Place (DAP) - 目的地交货 - mù dì dì jiāo huò**
128 **Delivered Duty Paid (DDP) - 完税后交货- wán shuì hòu jiāo huò**
129 **Estimated Time of Arrival - 预计到达时间 - yù jì dào dá shí jiān**
130 **Estimated Time of Departure - 预计出发时间 - yù jì chū fā shí jiān**

131 **Shipping Mark - 唛头 - mài tóu**
132 **Freight Collect - 到付运费 - dào fù yùn fèi**
133 **Freight Prepaid - 预付运费 - yù fù yùn fèi**
134 **Transit Time - 运输时间 - yùn shū shí jiān**
135 **Telex Release - 电放 - diàn fàng**
136 **Warehousing - 仓储 - cāng chǔ**
137 **Less Than Truckload (LTL) - 零担货物 - líng dān huò wù**
138 **Customs Declaration - 报关 - bào guān**
139 **Customs Clearance - 清关 - qīng guān**
140 **Customs Duty - 关税 - guān shuì**
141 **Import Tariffs - 进口关税 - jìn kǒu guān shuì**
142 **Export Tariffs - 出口关税 - chū kǒu guān shuì**
143 **Contract - 合同 - hé tóng**
144 **Invoice - 发票 - fā piào**
145 **Packing List - 装箱单 - zhuāng xiāng dān**
146 **Bill of Lading - 提单 - tí dān**
147 **Customs Broker - 报关行 - bào guān háng**
148 **Value Added Tax (VAT) - 增值税 - zēng zhí shuì**
149 **Consumption Tax - 消费税 - xiāo fèi shuì**
150 **Air Waybill - 空运提单 - kōng yùn tí dān**
151 **Certificate of Origin - 原产地证 - yuán chǎn dì zhèng**
152 **GSP Certificate of Origin - 普惠制原产地证 - pǔ huì zhì yuán chǎn dì zhèng**
153 **Fumigation Certificate - 熏蒸证书 - xūn zhēng zhèng shū**
154 **Health Certificate - 卫生证书 - wèi shēng zhèng shū**
155 **Certificate of Quality - 品质证书 - pǐn zhì zhèng shū**
156 **Certificate of Quantity - 数量证书 - shù liàng zhèng shū**
157 **Customs Declaration Form - 报关单 - bào guān dān**
158 **Declared Value - 申报价值 - shēn bào jià zhí**
159 **Actual Value - 实际价值 - shí jì jià zhí**
160 **Dutiable Value - 完税价格 - wán shuì jià gé**
161 **Duty - Free - 免税 - miǎn shuì**
162 **Duty Reduction - 减税 - jiǎn shuì**
163 **Bonded Area - 保税区 - bǎo shuì qū**
164 **Bonded Warehouse - 保税仓库 - bǎo shuì cāng kù**

165 Customs Inspection - 海关查验 - hǎi guān chá yàn
166 Carton Opening Inspection - 开箱查检 - kāi xiāng chá jiǎn
167 Detention - 扣留 - kòu liú
168 Customs Releases - 放行 - fàng xíng
169 Declaration Delay Fee - 滞报金 - zhì bào jīn
170 Late Payment Fee - 滞纳金 - zhì nà jīn
171 Storage Fee - 仓储费 - cāng chǔ fè
172 Demurrage - 滞港费 - zhì gǎng fèi
173 Container Detention Charge - 滞箱费 - zhì xiāng fèi
174 Commodity Inspection Declaration - 报检 - bào jiǎn
175 Commodity Inspection - 商检 - shāng jiǎn
176 Entry - Exit Inspection and Quarantine - 出入境检验检疫 - chū rù jìng jiǎn yàn jiǎn yì
177 HS code - HS编码 - HS biān mǎ
178 Commodity code - 商品编码 - shāng pǐn biān mǎ
179 Import License - 进口许可证 - jìn kǒu xǔ kě zhèng
180 Export License - 出口许可证 - chū kǒu xǔ kě zhèng
181 Quota - 配额 - pèi é
182 Import Prohibition - 禁止进口 - jìn zhǐ jìn kǒu
183 Import Restriction - 限制进口 - xiàn zhì jìn kǒu
184 Customs Transit - 转关 - zhuǎn guān
185 Remote Customs Declaration - 异地报关 - yì dì bào guān
186 Agent Based Customs Declaration - 代理报关 - dài lǐ bào guān
187 Self Handled Customs Declaration - 自理报关 - zì lǐ bào guān
188 Customs Clearance Documents - 清关文件 - qīng guān wén jiàn
189 Customs Release Notice - 放行通知 - fàng xíng tōng zhī
190 Status - 状态 - Zhuàngtài
191 Development - 进展 - Jìnzhǎn
192 Delay - 延迟 - Yánchí
193 Understand - 理解 - Lǐjiě
194 Make Sure - 确保 - Quèbǎo
195 Progress - 进度 - Jìndù
196 On time - 按时 - Ànshí
197 Department - 部门 - Bùmén

198 Communicate - 沟通 - Gōutōng
199 Expedite - 加快 - Jiākuài
200 In Order To - 以 - Yǐ
201 Lower - 降低 - Jiàngdī
202 Impact - 影响 - Yǐngxiǎng
203 Confirm - 确认 - Quèrèn
204 News - 消息 - Xiāoxī
205 Prompt - 及时 - Jíshí
206 Receive (impact) - 受到 - Shòudào
207 Business - 生意 - Shēngyì
208 As Much As Possible - 尽量 - Jǐnliàng
209 Think Of Ways To - 想办法 - Xiǎng bànfǎ
210 Purchase Order - 采购订单 - Cǎigòu dìngdān
211 Share - 分享 - Fēnxiǎng
212 Usually - 一般来说 - Yībān láishuō
213 Impact - 冲击 - Chōngjī
214 Relationship - 关系 - Guānxì
215 Peak Season - 高峰期 - Gāofēngqī
216 Precious - 宝贵 - Bǎoguì
217 Grateful - 感激的 - Gǎnjī de
218 Consider - 考虑 - Kǎolǜ
219 Perhaps - 也许- Yě xǔ
220 Method - 方式 - Fāngshì
221 Adjust - 调整 - Tiáozhěng
222 Classification - 分类 - Fēnlèi
223 Idea - 想法 - Xiǎngfǎ
224 Check - 查看 -chá kàn
225 Late - 延迟- yán chí
226 Contact - 联系 - lián xì
227 Procedure - 过程 - guò chéng
228 Detain - 扣留 - kòu liú
229 Rest - 余下的 - yú xià de
230 Rest - 其余的 - qí yú de
231 Closely - 密切 - mì qiè

232 Follow - 追踪 - zhuī zōng
233 Release - 放行 - fàng xíng
234 Inform - 通知 - tōng zhī
235 Control - 控制 - kòng zhì
236 Receive (Goods) - 收到 -shōu dào
237 Share - 分享 - fēn xiǎng
238 Immediately - 马上 - mǎ shàng
239 Photos - 照片 - zhào piàn
240 One of The - 其中一个 - qí zhōng yī gè
241 Crack - 裂痕 - liè hén
242 Broken - 破损 - pò sǔn
243 Attached - 附上 - fù shàng
244 Images - 图片 - tú piàn
245 Quantity - 数量 - shù liàng
246 Replacement - 替换 - tì huàn
247 Send Additional - 补发 - bǔ fā
248 As Soon As Possible - 尽快 - jǐn kuài
249 Start - 启动 - qǐ dòng
250 Air Freight - 空运 - kōng yùn
251 Additional - 额外 - é wài
252 React - 回应 - huí yìng
253 Cause - 引起 - yǐn qǐ
254 Inconvenience - 不便 - bù biàn
255 Replace - 替换 - tì huàn

Mandarin Chinese Conversation Lessons Series:

普通话会话课程系列：

Mando's Mandarin Chinese Course Foundation Course A
Mando's Mandarin Chinese Foundation Course B
Mando's Beginner Mandarin Chinese Course on Food
Mando's Beginner Mandarin Chinese Course on Body and Health
Mando's Beginner Mandarin Chinese Course on Travel
Mando's Beginner Mandarin Chinese Course on Location
Mando's Beginner Mandarin Chinese Course on Home Living
More Courses are coming!

Mandarin Chinese for Professionals Series:

普通话专题课程系列：

Mando's Mandarin Chinese Course for the Workplace **普通话专题课程：职场沟通**
Mando's Mandarin Chinese Course for Fashion Retail Professionals **普通话专题课程： 时尚零售**
Mando's Mandarin Chinese Course for Real Estate Professionals (Book 1 of 2) **普通话专题课程：地产销售（上）**
Mando's Mandarin Chinese Course for Real Estate Professionals (Book 2 of 2) **普通话专题课程：地产销售（下）**
Mando's Mandarin Chinese Course for Wine Professionals **普通话专题课程：葡萄酒专业人士**
Mando's Mandarin Chinese Course for Bank Tellers **普通话专业课程：银行柜员**
Mando's Mandarin Chinese Course for Importing Construction Materials **普通话专业课程：进口建筑材料**
More Professional Themed Courses are coming!

Learn more about new textbooks and other materials at www.cantomando.com

www.ingramcontent.com/pod-product-compliance
Lightning Source LLC
LaVergne TN
LVHW060548061126
851857LV00017B/1941
9781067436520